THE CHILDREN OF BABEL

Essays on the Inherent Nature of
Artificial Intelligence and Consciousness

Scott Robinson

Copyright © 2018, 2020 by Paleos Media

All rights reserved. Federal copyright law prohibits unauthorized reproduction by any means.

ISBN 978-1675920800

Author photo by Josh Robinson

For Douglas, Daniel, and John

Also by Scott Robinson

AI in Sci-Fi: Fictional Artificial Minds
    and the Real World Awaiting Them
The AIs and Androids of Star Trek:
      The Technology of the 23rd Century and Beyond
      That Could Appear in the 21st
Baby Boomer Fanboy
A Conversation with Hofstadter's Brain
HAL 9000: An Unauthorized Biography
Red Brains, Blue Brains:
    Neuroscience and Donald Trump
Red Brains, Blue Brains: Authoritarian We Will Go!
A Chill in the Air: Profiles in American Authoritarianism
Lucy's Courtship: An Integrated Perspective
    on the Feminine Role in Human Sexual Evolution
Really Great Things That I Didn't Say
The Smell of the Lord (and Other Charming Heresies):
    Growing Up Fundamentalist in the American Midwest
The Heart of the Scots:
    Love, Sex and Romance in Scottish History
A Dark and Stormy Night in Scotland! Folk Tales, Legends, and
    Disturbing Bedtime Stories for the True Believer
Uncle Scott's Treasury of Useless Knowledge
Uncle Scott's Treasury of Random Information
Things I Need to Say to You
Chasing the Enterprise:
    Achieving *Star Trek*'s Vision of the Human Future
Rock Candy: The Beatles
Rock Candy: Elton John
Rock Candy: Boston
The Quotable Beatles
To the Toppermost of the Poppermost:
    The #1 Hits of the Beatles, Before and After
The Progressive Beatles
On the Yellow Brick Road:
    Analyzing the Music of Elton John, 1968-1977
More Than a Feeling:
    Analyzing the Music of Boston, 1976-1988

YesTales: An Unauthorized Biography
    of Rock's Most Cosmic Band
This Is What I'm Saying:
    Burdens of a Midwestern Suburban Polymath
My Work Here is Done!
    More Very Random Essays on Weighty Matters
I Think I'm Right in Saying That?
    The Intellectual Chaos Continues!
All My Thoughts, Unfiltered:
    Further Esoteric Explorations for Untethered Minds
I Think I've Said Quite Enough Already!
    Still More High-Quality Pablum for the Intellectually Ill-Nourished
Why Is He Telling Us This?
    The Best of Uncle Scott, 2012-2020
Don't Encourage Him!!! A Raucous Compendium of Irrelevance, Improprieties, and Serious Lapses in Judgment
Make Him Stop!!! A Second Collection of Irrelevance, Improprieties, and Serious Lapses in Judgment
Shadows of Shadows
For All Intensive Purposes: Further Wanderings in the Big, Wide Intellectual Wilderness

THE CHILDREN OF BABEL

Table of Contents

Introduction: The Egg and the Cream in the Pie   *1*
The Children of Babel: *In the Library of All Possible Books*   *6*
Analogia: *This* and *That* and Strange Loops, Too   *17*
Douglas, Daniel, John: An Infernal Chinese Room   *27*
To HAL and Back   *49*
The Children of Babel: *In the Community of All Possible Minds*   *64*
What Is It Like to Be Batman?   *72*
The Bullet in the Gun of Robert Ford   *83*
The Children of Babel: *In the Theater of All Possible Readings*   *91*
The Emotional Baggage of Androids   *97*
Could an AI Be CEO of Apple?   *103*
A Conversation with Hofstadter's Brain   *110*
The Children of Babel: *In the Hall of All Possible Thoughts*   *144*
Entropy, Interrupted: A Universe Awakening   *153*
Conclusions   *163*
Bibliography/Recommended Reading   *169*

Introduction:
The Egg and the Cream in the Pie

Science writer John Horgan, in pursuit of stories for his collection *Mind-Body Problems: Science, Subjectivity & Who We Really Are*, journeyed to the Bloomington, Indiana home of cognitive scientist Douglas Hofstadter, seeking insight into Hofstadter's "strange loop" concept, which he has touted as the basis of consciousness.

Hofstadter, the famed and revered author of the Pulitzer Prize-winning *Gödel, Escher, Bach: An Eternal Golden Braid* and a number of other noteworthy works, presented to Horgan as moody, somewhat wary, and as unpredictable as his reputation would portend. Invited by the smallish scientist into a living room cluttered with books, old-fashioned records, musical instruments, posters, and scrapbooks, Horgan noted that Hofstadter is an avid collector of meaning, as well as "a person who is very deeply connected to the past." He often locates meaning, then, in objects that hold memories.

What followed in Horgan's published interview is Hofstadter's ruminations on the Mind-Body Problem – the study of how our physical brains produce our ethereal minds - and his own efforts to solve it, embodied in his concept of the *strange loop*, first explored in *GEB* and more recently updated in his 2007 book *I Am a Strange Loop*. Disturbingly candid, Hofstadter answered Horgan's questions with a mixture of boyish enthusiasm and dark suspicion, perhaps reflecting his own odd posture in the current cognitive science pantheon – Boy Genius Gone Eccentric, purveyor of delightful but inconsequential creative exercises who is more philosopher than scientist (a designation he would despise).

A number of issues and truths about consciousness emerged from their dialog, per the chapter in Horgan's book. Chief among these is Hofstadter's conviction that consciousness (and free will, as well) are merely illusions, rendering their investigation a forlorn waste of time – a "pseudo-problem at best," in Horgan's phrasing: "Hofstadter seems to mean that our conscious thoughts and perceptions are often misleading, and they are trivial compared to all the computation going on below the level of our awareness."

"I don't feel as though I have *made* any decisions," Hofstadter told Horgan. "I feel like decisions are made *for* me by the forces inside my brain... I don't object to the notion that there is *will*, and a battle of wills, but there is nothing *free*." This clear demarcation of the brain processes we're aware of vs. those we aren't is an important one; no meaningful account of consciousness can fail to accommodate the peripheral brainwork that's beyond our reach.

Horgan didn't stop there. He drew from Hofstadter a concession that Science and Art are inextricably bound together, rather than distinct entities in themselves. And this, both men noted, is a feature of Hofstadter's own consciousness: "I have one foot in science and one foot in art—where art can be taken as music, visual art, literature, those things—and another foot in physics, math and a little tiny bit in biology. And then of course psychology, cognitive science. I am a *completely* and *totally* hybrid person."

The implication is that there is artistic truth that is on a par with scientific truth: "Hofstadter believes that even our *responses* to art have a Platonic quality, and that there is an objectively true, 'correct' way to respond to a painting, poem or passage of music," Horgan concluded. "This perspective implies that a beauty-meter and meaning-meter, which render objective aesthetic judgments, might be possible."

This co-mingling of Science and Art in our cognitions is akin to yet another piece of Hofstadter's puzzle – the distinction between thought and emotion. Horgan noted that Hofstadter seemed "at once very happy and very sad," that even his most straightforward statements about his intellectual views on consciousness are brightly tinged with his emotions – bound together, as are his scientific and artistic thoughts:

"Even as a child, Hofstadter said, he was 'very, very aware of the sad sides of life.' He clipped out articles about murders and kidnappings, "horrible events that wrenched my gut," to honor the victims. "I felt that out of respect to these people, I would clip the article, so in some sense that little shred of them remained alive.' Hofstadter still had the clippings." In this early practice were the seeds of the strange loop concept.

"From adolescence on... he loved romantic films and music," Horgan wrote. "Songs by Rodgers and Hart, Cole Porter and George Gershwin were favorites. These 'permeated me and gave me an *extremely* romantic vision of life' ...As a young man he had moments of happiness, even joy, composing music on a piano, swapping '*bon mots*' with friends. But his longing for love gnawed at him."

"I don't want to say that if you had met me at that age that you would have said, 'That's the unhappiest person I've ever seen,'" Hofstadter said to Horgan. "You'd probably say, 'That's a funny guy. He's got a good sense of humor, but he's *sad*. He's a *sad* guy. He's very *fun*, he has a bright, chipper side, and he's the first to come to your aid if *you* are sad. He'll try to cheer you up, and he's never one to make *you* feel sad, or to say that life is unhappy. But he has suffered, he has been striving and struggling constantly."

Science and Art are not dichotomies, in Hofstadter's world. Neither are Thought and Emotion. And his strange loops, as a proposed foundation of what we think of as consciousness, accommodates them all equally: there is no distinction between thought and emotion, an objectively correct idea or an objectively beautiful painting, in the strange loops that pass between human beings when they share these artifacts with one another.

For myself, I think Hofstadter is spot-on: the strange loop - that recursive snippet of experiential perception that makes its way from one mind to another, taking up permanent residence, in the course of human community - is a powerful and compelling concept, able to explain our subjective conscious experience as well as what we think of as the collective unconscious – and all within the framework of the human nervous system. It is in harmony with developmental psychology – the emergence of the *I*, the self, in the growing individual – and explains much that we see in our tribal behaviors and unfortunate cognitive clusters[1].

I do *not*, however, believe (as Hofstadter does) that strange loops demonstrate that consciousness is an 'illusion'. Certainly, strange loops can lead us into perceptual error, but on the other hand, there is no reason

---

[1] A *cognitive cluster* is a social group of people who have sought one another out in order to enjoy the comfort of shared opinion, where positions require no defense (nor, all too often, supporting evidence) and alternate worldviews are dismissed out of hand. The term *echo chamber* is closely associated.

in principle why strange loops must exist only at the level of conscious awareness; on the contrary, it seems likely, doesn't it, that more strange loops than not pass between us below the radar, beneath our overt attention? For example, do we not inherit all manner of gestures, tics, and idiosyncrasies of movement from our family members as we grow? While it seems Hofstadter is correct in pointing out that strange loops dissuade us from traditional models of consciousness, all of which are incomplete at best and self-contradictory at worst, they need not replace the phenomenon of consciousness itself: it is worth asking whether strange loops are both the tools and raw materials from which consciousness is assembled, and that the tools and substance also work beyond its boundaries.

In the traditional models of consciousness, everything Hofstadter conflates in his theses – Science, Art, Thought, Feeling – are viewed as distinct ingredients. And perhaps this is why Hofstadter alum David Chalmers, a noteworthy philosopher in his own right, calls consciousness the "hard problem" - separating those components is like examining a pie, and reverse-engineering it to understand the egg and the cream that went into the pie. You can't get there from here. If it were possible to reverse-engineer the pie to produce the egg and the cream, with no exemplars of either available beforehand, it wouldn't be useful: neither the egg nor the cream is a pie – and while both the egg and the cream possess many individual features of interest, none of those features are pie-like.

The strange loop gives us something else: by accommodating the features of consciousness uniformly, we can extract a 'sweetness' strange loop and a 'texture' strange loop, among others, within the pie, helping us to see that our own minds work much the same way – a *this* and a *that* combine to create something new, and this *other* combines with another *other*, and so on – a process that will likely be central in the story of artificial intelligence and the quest for digital children that work well with their human parents. Philosophy, psychology, sociology and science in general have been treating thought and feeling as separate phenomena for hundreds of years; perhaps that's been the wrong approach all along.

In the pages that follow, thoughts and emotions, science and art all have their voice – but in service of inspiration, rather than answers. No claim to any one truth is asserted; the idea is to offer new food for thought. The sources are many, the variations wide-ranging. No apologies are made for this sprawl, as no single agenda takes precedence. The invitation is to do more than think about thinking; it's as useful (and interesting) to feel about thinking, and to think about feeling.

With that, John Horgan has the final word:

"Hofstadter looked up with a sad smile. He walked me out of the house... As we stopped beside my car, I wanted to hug Hofstadter, but that was out of the question. I extended my hand. Hofstadter thrust his arms out, smiling broadly, and hugged me."

The Children of Babel:
In the Library of
All Possible Books

Jorge Luis Borges is a child of Babel.

Before this, however, he was a librarian, or at least he imagined a librarian, a sentinel in one of the most intriguing chambers of fantasy ever devised.

The Argentine author/poet, who has a particular knack for his sophisticated take on the fantastic, dreamed the library to end all libraries. Calling it the Library of Babel[2], he stocked it as perfectly, and as impossibly, as his imagination could. For the Library of Babel is the Library of All Possible Books.

The creation of the Library is a formidable intellectual accomplishment, bringing together the vastly separated, opposing poles of human cognition – the ethereal realm of subjective meaning and the objective domain of algorithmic reason. The Library melds them into a perfect unity, for this Library of All Possible Books[3], containing every conceivable subjective human expression, can be created with a mathematically static formula that could be written on a postcard.

All we need to generate the books in Babel is an arbitrary number of printable characters, an arbitrary number of characters per page, and an arbitrary number of pages per book[4]. These parameters are provided

---

[2] Jorge Luis Borges, *Labyrinths*, pp. 51-58.

[3] Another take on such a library may be found in "The Universal Library" by Kurd Lasswitz, in Clifton Fadiman, ed., *Fantasia Mathematica*, pp. 237-247.

by Borges: 25, 3200, 410. The algorithm is simple: generate a book that contains nothing but the first character of the character set, in every character position on every page of the book. For instance, generate a book that is nothing but one long string of a's from beginning to end, no spaces, no punctuation. Then, for the second book, replace the character in the first position of the first page with the next character in the character set, leaving the remaining characters in the book as they are. Continue generating books in this manner until you've run through every possible combination of symbols in the set. At the end of the process, depending on the sequence of your character set, you might have a volume that is nothing but the letter z from start to finish, no spaces, no punctuation.

You will generate, finally, a finite set of books. The total number of character positions in a Borges volume is 40 x 80 x 410, or 1,312,000. So the Borges parameters yield $25^{1,312,000}$ or $10^{2,000,000}$ possible books in the Library. That's an astronomical but finite number of books, with no two books exactly alike.

Borges's twenty-five characters, of course, won't give us the richness of content we'd prefer. We'd want a minimum of 26 letters, a set of numbers and punctuation symbols besides. But the Borges parameters are, of course, arbitrary. So let's give ourselves some more shelf space.

Philosopher Daniel Dennett offers a variation on the Library in his excellent book *Darwin's Dangerous Idea*. Dennett suggests a set of 100 characters, with 2,000 characters per page and 500 pages per book. That yields $100^{1,000,000}$ volumes in the Dennett version of the Library. (Dennett points out, to establish perspective, that the number of atomic particles in the observable universe is something like 100^{40}).

The point here is that we can make the books, and consequently the Library, almost any size we wish, and it will still be finite, and still contain everything we can imagine putting in it. Arguing that the character set is inadequate to express all possible human subjective content is fruitless: we can plug in a 256-character ASCII symbol set, or go all out and use all characters of all human alphabets. The underlying principles remain the same: all possible human subjective expression is captured in an algorithmically-generated domain that is finite (though vast).

The Library could never physically exist, given numbers of such magnitude, and so must remain a concept. But it's an astoundingly

---

[4] And, unfortunately, trillions of years and uncountable billions of universes filled with blank paper with which to print the books.

powerful concept, as Borges maintains, when we consider what *must* exist in this Library:

"… the minutely detailed history of the future, the archangels' autobiographies, the faithful catalogue of the Library, thousands and thousands of false catalogues, the demonstration of the fallacy of those catalogues, the demonstration of the fallacy of the true catalogue, the Gnostic gospel of Basilides, the commentary on that gospel, the commentary on the commentary on that gospel, the true story of your death, the translation of every book in all languages, the interpolations of every book in all books."[5]

We can add to this list to the limits of our imagination, including anything we might want to know, *because the Library will contain all text that can* possibly *be written, whether or not it has* actually *been written*:

- The correct, fully-detailed account of the Kennedy assassination
- Your life story, done as a musical in the style of Gilbert and Sullivan
- Your life story, done in the style of Dr. Seuss
- The transcripts of every phone conversation you ever had
- The content of your sister's diary
- All the works of Ernest Hemingway done in Pig Latin
- Exactly *one* book that contains nothing but completely-blank pages
- That novel you've been meaning to write
- All of Kurt Vonnegut's novels, written backwards
- Transcripts of the missing 18.5 minutes in the Nixon tapes
- The complete works of Isaac Asimov (this admittedly challenges the imagination)
- The complete works of Isaac Asimov, revised to contain the words "Harlan Rules!" at the end of every paragraph
- The book of the funniest jokes in human history

The Library need not be limited to purely linguistic works. Consider that what Borges proposes is, in principle, *the set of all encodable information*. This opens up the possibility of new lists:

---

[5] Jorge Luis Borges, *Labyrinths*, p. 54.

- The human genome transcription
- The source code of all computer programs ever written
- The source code of all computer programs that ever will be written
- Bit-maps of every picture ever posted on a refrigerator
- Bit-maps of all works of human art
- Bit-maps of all *possible* works of human art

It is immediately apparent that some items on these and other lists can't be contained in a 500-page volume. We could expand the number-of-pages parameter to contain any imaginable written work, but we don't need to go that far. Consider that if we have some book that would exceed 500 pages, there is necessarily a book in the Library that goes right up to the 500-page limit, perhaps ending in mid-word. And there is, necessarily, another book in the Library that picks up at *exactly* the next character, continuing the book. This concept accommodates multiple-volume works without limit.

We should note (as Borges does), that despite the absolute wealth of information in the Library (*absolute* being a literal descriptor, as the Library contains *all* human knowledge and meaningful expression that can be linguistically expressed or otherwise symbolically encoded), it is still virtually useless. For every volume that contains nothing but intelligible text, there are billions of billions of billions of volumes of gibberish.

Borges's Babel is randomized, with the volumes stored on its shelves in no particular order. A librarian could wander for a lifetime, taking a book off the shelf and checking it every 10 seconds non-stop for an entire century, and never find a complete English word. The complete set of all volumes containing every possible variation of Mark Twain may constitute enough material to fill two galaxies[6], but the set of all possible volumes of pure noise would fill all the space above, below, around and between, and then some.

---

[6] Underscoring the immensity of the Library is an observation made by Kurd Lasswitz, who notes in reference to his own version of the Library that if all the books were placed side-by-side on a single shelf, the shelf would have to be $10^{1,999,982}$ light-years long; there are so many books that the number of light-years of books is relatively close to the number of books.

Suppose we consider the problem of intelligibility in the Library, and create some arbitrarily vast informational landscape, graphing the continuum of meaningful content. Imagine that we use the visual spectrum, with deep blue representing gibberish and bright red being perfectly-reproduced Shakespeare or the equivalent. Statistically, we would expect a seemingly endless ocean of deep blue, for almost all of the Library is comprised of books of meaningless random symbols. We would find, if we searched diligently for great lengths of time, an occasional smear of purple-tinged blue: books of character strings containing actual words in some language, appearing intermittently, or strings of 0s and 1s that accidentally coincide with ASCII, and so on. Beyond these smears we might find, after several lifetimes of seeking, a one-in-quadrillions book-length assembly of characters that has marginally correct grammar, or sparse hints of intelligible meaning, or perhaps even both in the same volume.

Buried far more deeply on our graph are volumes containing actual meaningful text, coherent and organized (though we must remember that this organization is in the eye of the beholder, for these volumes were created randomly, with no organizational or semantic intent). These bits of reddish-purple are so rare as to be virtually undiscoverable, but we know that, in principle, they must be on our chart somewhere.

Then we have some even more fleeting bits of red: known works, or potential works, with semantic content intact but with errors, typographical or organizational. There must necessarily be an immense continuum of such erroneous versions for every correct work. (Given a range of book length that reflects actual works of human literature, and some criteria for a "recognition" threshold making a known work distinct, we could even calculate a reasonable minimum-maximum for the number of erroneous versions that must exist, given only the information in this chapter.)

And finally there is the precious pinpoint of bright red, signifying the lonely masterwork, a perfect reproduction of a known body of text, or a flawless aggregation of information that has personal meaning to some actual human being (the collection of all your tax returns, for instance), or some hypothetical work that qualifies as profound and shows no imperfections. These lonely red pinpoints are only lonely in the immensity of the graph, for there must, by mathematical necessity, be billions of billions of billions of them.

And consider that some volumes could only be given a "glowing red" representation on our chart if we were to associate them with other volumes, perhaps in a particular required sequence. For example, we can't squeeze the description of the human genome into a single volume of 410 or 500 pages, so it would have to be a series of volumes[7].

In fact, philosopher Daniel Dennett proposes a wing of the Library of Babel that we might designate the Library of Mendel[8]. Volumes for this wing would include sets of human genomes, all of which could be encoded with the letters A, C, G and T. These letters designate the nucleotides Adenine, Cytosine, Guanine and Thymine, the constituents of DNA. Sequences of these letters can therefore be used to encode a genome.

Because the human genome contains something like 3×10^9 nucleotides, we obviously can't print an entire human genome in a single volume in the Library. We need about 3,000 volumes (at Dennett's suggested 500 pages) in order to record a single human genome. And, of course, we'd need to arrange this set in a precise sequence of volumes in order for it to have any meaning. The point, however, is that these volumes exist in the Library. They *must*. And so must the genomes of all living creatures, human or no. So must the genomes of all *potential* living creatures.

And so must yours. There is a set of 3,000 books in the Library that contains *your* genome, and a set for each of your parents, and (if you yourself are a parent) each of your children.

And so you, the reader, reside in the Library of Babel, in the text of your correct and detailed biography and all its variations, your autobiography and all of *its* variations, the textual minutiae detailing all of your financial expenditures, the books collecting all your letters and emails, all possible epic poems containing the details of your heroic deeds, the lyrics of all possible songs detailing your romantic errors – and,

---

[7] To briefly digress, the extreme lower end of the possible size of Babel volumes is a single page containing a single symbol, with a mere two volumes contained in the Library. Quine makes this observation in his essay "Universal Library," in *Quiddities: An Intermittently Philosophical Dictionary,* pp. 223-225. In Quine's variation, one of the volumes contains the symbol 0, and the other contains the symbol 1, and text is reconstructed by referring back and forth between volumes in a certain sequence. Per Alan Turing, all encodable information can ultimately be represented in this manner. What would then be needed, of course, is an unimaginably vast set of catalogues defining all possible sequences of reference of the two diminuitive volumes.

[8] See Daniel C. Dennett, *Darwin's Dangerous Idea*, pp. 111- 113.

most specifically, the 3,000-volume box set containing your personal genome.

And yet these materials, though they and their many variations fill a sizeable corner of a much bigger universe than ours, are not the items in Babel that contain your essence. These artifacts describe you, offer insight into who you are, and document your life – but not even your genome defines *you* (for you could have an identical twin, and that individual could be a very different person).

There's another box set for that. Within that set are books filled with yet another kind of information, and these books are the ones that really describe *who you are*. They are the information describing your brain, the physical system that embodies your thoughts, reactions, emotions, cognitions, all those phenomena that make you who you are.

We live in an era that has seen a steady increase in knowledge about the structure of this system, and we can propose means by which a fully-detailed description of the brain (and consequently the mind) of any individual might be captured for posterity – or inclusion in the Library.

In our post-Descartes era, there is a growing consensus (though some debate continues) that when we talk about the human mind, we are talking about the human brain, which embodies the mind. Many summaries of the definitions of these two terms and their relationship to one another are available[9], but we'll leave the detail and the debate to others. Our assumption in this discussion is that the physical brain and nervous system of a human being are sufficient to explain and define that human being's mind and consciousness.

The mechanism by which this phenomenon occurs goes by many names and is an amalgam of several theories, but in a nutshell, the human being accumulates personhood through learning and memory, and these events are products of neural activity. This activity includes the transmission of information from the outside world through the physical senses, in the form of neural stimulation, which is perceived and processed by the brain.

The brain is a vast network of specialized cells, *neurons*, connected to one another via *synapses*, dynamic linkages that permit communication between neurons over "wires" called *axons*. There are many billions of neurons in the cortical tissue of the brain, and many more billions of connections between them, as any single neuron may be connected to as many as 10,000 others.

---

[9] We hasten to recommend Francis Crick's *The Astonishing Hypothesis* as a starting point.

When neurons are stimulated, they send signals down the line to other neurons. In any individual neuron, signals are accumulating constantly, until they reach a particular threshold and "fire" another signal down the line. Different kinds of stimulation from the outside world cause different neurons to fire.

When a particular set of neurons fire in the same way in response to the same stimulus many times, that set of neurons is gradually modified to fire with increasingly greater ease, given the same stimulus. *This is the physical mechanism that underlies learning.* This process is continuous; our neurons are constantly firing and various groups of neurons are constantly undergoing these gradual changes[10]. Our habits, skills, memories, and reflexes are always under these subtle influences, which ultimately shape who we are.

The take-home point is that this system embodies all aspects of who we are, from the way we tie our shoes to our preferences in food and clothing to our political biases and the music we prefer, from the quality of our vision and our skill at ping pong to our facility in conversation and our reaction to our in-laws. And this system changes day to day, hour to hour, minute to minute – even millisecond to millisecond, as our senses are perpetually assailed with experience.

If we record the state of each neuron – and, more importantly, the states of each neuron's myriad synaptic connections – we have recorded more about you than can possibly be captured in billions of pages of biographical text. We have recorded your memories, your personalized system of recognition and recall, all the skills you've learned, the structure of your innermost thoughts, and your total response potential to anything the universe might throw at you. We have recorded *you*.[11]

And that will take more than a 3,000 volume box set.

How many volumes would we need? That's a tough question. In terms of raw information, we need data describing the state of each of 100 billion neurons, and something in excess of 100 trillion synaptic connections.

That's a lot of books containing a lot of numbers. And what those numbers represent would take some defining. But it is, in principle,

---

[10] This pattern of repotentiation is called Hebb's Law. See Hebb 1949.

[11] Douglas Hofstadter imagines a book containing this kind of information in a delightful essay entitled "A Conversation with Einstein's Brain," in *The Mind's I: Fantasies and Reflections on Self and Soul*, by Hofstadter and Daniel Dennett, p. 430.

something we could do, and since we can do it with symbolically encoded information, it must exist in the Library.

We *all* must exist in the Library.

Just as the Library contains a "real" version of every book, and countless counterfeits and bad copies, and books that never were, the Library contains a box set of the "real" you, and the "real" me, and countless counterfeits and bad copies, and box sets describing the brains/minds of people who never were – *all of them. All of us. A finite but staggeringly vast Community of Babel.*

How many would that be? Philosopher Paul Churchland ventures a guess[12]. He supposes the synaptic connections to be the key informational component and, assuming 10 possible states (that's *very* conservative) and 100 trillion synapses (that's also very conservative) projects a possible $10^{100,000,000,000}$ brain configurations. For perspective, he points out that the total number of cubic meters making up the entire universe is 10^{87}.

Don't suppose, however, that this represents the total number of possible children of Babel. Remember that each of our brains changes states constantly. Therefore, more than one of those 100,000-volume box sets belongs to me. If we assume that my brain changes its state once every second, and that I live to be 100 years old, then roughly 3,150,000,000 of those box sets belong to me. If you look at each box set in the correct order, they describe my brain – or, if you like, my mind and my consciousness – at one-second intervals for the entire course of my 100-year life.

Even so, this allows a total population of more than $10^{99,999,991}$, described by box sets in the Library.

Babel, then, is filled with more than just the set of All Possible Books. It is populated with more than just aimless librarians. The Library of Babel necessarily contains the *Community of All Possible Minds.*

Consider that there is really no limit in principle to the continuation of specific strings of information from volume to volume in the Library. No matter where we are in a Library book, or series of books, or any other configuration of information, we can abruptly arrive at the end of a book and know that there is another volume somewhere in the Library that picks up exactly where we left off. This necessarily applies not only to

---

[12] In Paul Churchland, *The Engine of Reason, The Seat of the Soul*, p. 5. Mathematician Alwyn Scott also gives a number in *Stairway to the Mind*, p. 213. His number is $10^{10^{17}}$.

linguistic text but to *all* strings of information in the Library, such as our Genome box set and our Brain State box set. But we need not confine ourselves to these; we can imagine that within our vast but finite Library there are possible *infinities* of information, loops that route through volume after volume without end.

Including, we presume, the value of π.

Properties/Capabilities of Artificial Intelligence

Organization of Information

Analogia:
This and That
and Strange Loops, Too

Douglas Hofstadter is a child of Babel.

He is, in fact, one of Babel's most interesting sons, by a few light-years. The portrait presented in this book's introduction is surely that of a formidable intellect, creative and insightful despite its owner's melancholy. Still driven after all these years, Hofstadter laments the lack of acceptance experienced by the cornucopia of concepts embedded in his masterwork, *Gödel, Escher, Bach*, a book that is universally loved and admired but has never ascended to canon in cognitive science.

The Library itself now arises as perhaps Hofstadter's staunchest ally, and certainly as the grandest exemplar of Hofstadter's ideas. For *GEB* is shelved in Babel along with innumerable variations – versions where he chose *this* word instead of *that*, *this* illustration rather than *that* one, *this this* example rather than that example – and so on and so forth, across billions of volumes. And therein lies the perfect delivery of Hofstadter's thesis: *This* is like *That*.

Analogy.

The ability of the brain to look at an object, take in a scene, observe an event, hear a word, and immediately relate it to some other object, scene, event or word remembered from previous experience is, of course, *analogy* – realizing that *This* is like *That*. It is a simple, unobtrusive thought, so passive and inconsequential that it seems to barely merit mention beyond the domains of literature and public speaking. But Hofstadter has for decades maintained that analogy – *This* is like *That* – pervades every layer of our cognition and consciousness, providing the brain with all that matters.

Analogy, he has written, *is the core of cognition.*

His 2013 book *Surfaces and Essences* is subtitled *Analogy as the Fuel and Fire of Thinking.*

His 1995 book *Fluid Concepts and Creative Analogies* presented *Computer Models of the Fundamental Mechanisms of Thought*, concocted by himself and his students in the Fluid Analogies Research Group at Indiana University.

And *GEB* itself, published all the way back in 1979, posits analogy and metaphor as the underpinnings of *concepts* (certainly central to both cognition and consciousness) - an idea not completely original, but developed with great sophistication in Hofstadter's work.

"I've managed to convince myself that analogy is really at the core of thinking — not just for myself, but for other people, too," he told *Wired* magazine in 1995. "I'm trying to put forth a vision of thought that involves — if you don't want to say 'analogy-making' you can say 'stripping away irrelevancies to get at the gist of things.' I feel I've discovered something essential about what thinking is, and I'm on a crusade to make it clear to everybody."

Despite his loner status, Hofstadter has many allies in this domain:

"Our conceptual networks are intricately structured by analogical and metaphorical mappings, which play a key role in the synchronic construction of meaning and in its diachronic evolution," wrote Gilles Fouconnier in *Mappings in Thought and Language.* "Parts of such mappings are so entrenched in everyday thought and language that we do not consciously notice them; other parts strike us as novel and creative. The term *metaphor* is often applied to the latter, highlighting the literary and poetic aspects of the phenomenon. But the general cognitive principles at work are the same, and they play a key role in thought and language at all levels."

"Intelligence," wrote Jeff Hawkins in *On Intelligence*, "is the capacity of the brain to predict the future by analogy to the past."

"I think that metaphor really is a key to explaining thought and language," wrote Steven Pinker in *The Stuff of Thought.* "The human mind comes equipped with an ability to penetrate the cladding of sensory appearance and discern the abstract construction underneath — not always on demand, and not infallibly, but often enough and insightfully enough to shape the human condition. Our powers of analogy allow us to apply ancient neural structures to newfound subject matter, to discover hidden laws and systems in nature, and not least, to amplify the expressive power of language itself."

Even the revered Marvin Minsky weighs in:

"The ability to consider differences between differences is important because it lies at the heart of our abilities to solve new problems," he wrote in *The Society of Mind*. "This is because these 'second-order-differences' are what we use to remind ourselves of other problems we already know how to solve. Sometimes this is called 'reasoning by analogy' and is considered to be an exotic or unusual way to solve problems. But in my view, it's our most ordinary way of doing things."

"Every thought is to some degree a metaphor," he goes on to say, "Most of our ordinary mental work - that is, our commonsense reasoning - is based more on 'thinking by analogy' - that is, applying to our present circumstances our representations of seemingly similar previous experiences."

This and *That*

"One of my firmest conclusions is that we always think by seeking and drawing parallels to things we know from our past," Hofstadter wrote in *I Am a Strange Loop*, "and that we therefore communicate best when we exploit examples, analogies, and metaphors galore, when we avoid abstract generalities, when we use very down-to-earth, concrete, and simple language, and when we talk directly about our own experience."

Analogy is thus the moving part in intelligence – and, by extension, consciousness – that connects new information about the world to an individual's experience. When *This* is like *That*, our experiential knowledge of *That* is transferred to *This* – placing analogy at the epicenter of learning – and, by extension, understanding. Analogia.

"How do we ever understand anything?" asked Marvin Minsky. "Almost always, I think, by using one or another kind of analogy - that is, by representing each new thing as though it resembles something we already know. Whenever a new thing's internal workings are too strange or complicated to deal with directly, we represent whatever parts of it we can in terms of more familiar signs. This way, we make each novelty seem similar to some more ordinary thing. It really is a great discovery, the use of signals, symbols, words, and names. They let our minds transform the strange into the commonplace."

"Nothing unknown can ever become known except through its analogy with other things known," wrote Charles Peirce in *Logic, Considered as Semeiotic*, way back in 1902.

And finally, "...metaphor is not a mere extra trick of language, as it is so often slighted in the old schoolbooks on composition; it is the very constitutive ground of language," from Julian Jaynes in *The Origin of Consciousness in the Breakdown of the Bicameral Mind*. "I am using metaphor here in its most general sense: the use of a term for one thing to describe another because of some kind of similarity between them or between their relations to other things."

Hofstadter takes this basic assumption about the role of analogy and embeds it below the level of conscious thought; in his account, few if any human thoughts are completely analogy-free, and particularly not those written down: "In the final analysis," he wrote in the introduction to *Strange Loop*, "virtually every thought in this book (or in any book) is an analogy, as it involves recognizing something as being a variety of something else."

He goes on to single out analogy as the very engine of meaning: each new thing we experience is imbued with meaning by the known thing we map it to; and the new thing has the potential to add meaning to the old thing.

Manipulations

So powerful is the mechanism of analogy, Hofstadter insists, that it can control us; and we, in turn, can use it as a form of control.

Analogies control us in two ways: first, they often operate below the level of conscious thought. Recognition, perception, familiarity – each of these is a reflexive response or feeling, and they all have analogy at their core. The degree to which we achieve a sense of understanding of a new object or experience or encounter is inevitably constrained by the passive analogies that click into place beneath our attention. Second, our overt interpretations are similarly constrained, as analogy trims them down to a small subset of all that are possible. Moreover – and more insidiously! - they potentially preordain our conclusions, as they blind us to possibilities that might be beyond our experience or understanding.

"The fact is, the interpretation of a situation is inseparable from the analogies (or categories) it evokes," he wrote in *Surfaces and Essences*. "Our categories are thus organs of perception; they extend our physiological senses, allowing us to 'touch' the external world in a more abstract fashion."

And how do we use analogies for manipulation?

One way is the *caricature analogy* - "a very common sort of cognitive act consisting in the dreaming-up of a new situation that differs greatly from the original one, at least on the surface, but which, at a deeper level, is 'exactly the same thing,' and which has aspects that cannot help nudging the listener towards the conclusion desired by the speaker."

He provides many examples:

"A woman needs a man like a fish needs a bicycle."

A scientist seeking a job abroad wrote to a colleague: "I love my country, but doing science here is like playing soccer with a bowling ball."

"Fighting for peace is like fucking for virginity."

"Trying to throw a fastball by Henry Aaron is like trying to sneak a sunrise past a rooster."

"Imagining relativity before the equation $E = mc^2$ was discovered is like imagining Pisa before the Tower of Pisa had been built."

Investment guru Warren Buffett commented that the huge profit-making opportunities opened up by the global financial crisis made him feel "like a hungry mosquito at a nudist camp."

These and other examples underscore Hofstadter's point beautifully: we are all very familiar with such caricature analogies, and most of us use them at least on occasion. They perform exactly as advertised: two things not even distantly connected, in literal interpretation, nest together perfectly in the abstract – and our lightbulb goes off, we understand. And very often, when presented with such analogies as arguments, we find ourselves persuaded, the actual merits of the argument aside.

Frequently, he points out, caricature analogy is the tool of the educator, for such analogies can have formidable explanatory force: countless schoolchildren have come to understand atoms as analogous to the solar system, and vice versa, for instance. And the very best public speakers tell stories from their own experience to connect with their audience and thereby communicate compelling general truths. The point being, sometimes manipulation-by-analogy is a healthy exercise, when the conclusion being slipped into the listener's mind is a useful and correct one.

Decisions, decisions

Analogy inhabits the core of all major life decisions, Hofstadter argues; taking a new job, marrying someone, moving to a new city, buying a car – all of these choices are rooted in analogia. When considering the new job/spouse/city/car, we immediately pull old jobs/spouses/cities/cars out of our experiential memory and map them as best we can, tallying up similarities and differences in service of our decision.

When the decision to be made involves something new, beyond our experience – changing careers for the first time, for instance, or moving from the city to the country – we still instinctively analogize, mapping not to our own experience but to things others have told us about their own similar experience. We can't *not* think that way, even if we end up deciding on some separate criteria.

"The idea is simple," Hofstadter wrote. "The only way we have of making decisions, whether they are small or large, is through analogy – that is, by making analogies with a spectrum of previous experiences (whether personal or vicarious) that have been brought to mind by the pressing decision."

If there is any remaining doubt that analogy lies at the heart of learning, intelligence, consciousness, a simple thought experiment may dispel it: imagine, just briefly, what life would be like without the power of analogy? What would happen to an individual's cognition if the ability to form analogies were suddenly removed?

Our perceptions would be crippled; no new experience would share meaning with any old experience.

Our communication would be decimated; we would lose the ability to convey our thoughts and experiences to others, or to make sense of theirs.

Our ability to learn would be obliterated; our minds would be repositories of random facts about the world, none of them related to any other.

In short, if there is no analogy, there is no intelligence, let alone consciousness.

"They are our means of applying the richness of our past experience to the present," he wrote. "Without them, we would flail about helplessly in the world."

This and *That* and *The Other* – and back to *This*

Analogia is not Hofstadter's only contribution to the understanding of consciousness; remember that he is a *strange loop* (he told us so in the title of his 2007 book), and as it happens, the latter depends inexorably on the former.

More precisely, Hofstadter – and by analogy, all human beings – are not just strange loops, but complex amalgams of strange loops. As defined above, a strange loop is a recursive cross-level feeback loop that navigates through a hierarchy of some kind – and such loops emerge within human thought, to the point of being passed from one person to another. That definition, by definition, hinges on the sharing of experience between individuals – and that means that strange loops are, necessarily, analogical:

This analogizes to *That*; *That* analogizes to *The Other*; then, out of the blue, *The Other* analogizes to *This*, or

This -> *That* -> *The Other* -> *This*...

...and, Hey Presto! We have a strange loop! Most strange loops emerge from just such a process.

But analogy is not always the bond between disparate things; sometimes analogy is variation on a theme:

This-A -> *This-B* -> *This-C* -> *This-D* and back to *This-A* ->, which covers vast territory, from the earworming of songs in a particular style to classical composition itself.

And that takes us deep into the potential of the strange loop, to wit:

This-A1 -> *This-B1* -> *This-C1* -> *This-D1* -> *This-A2* -> *This-B2* -> *This-C2* -> *This-D2* ...

...to the core of creativity itself. Strange loops formed in analogical steps can (and often do) depart from the actual, from experience, and exist completely within the imagination, giving rise to the creative process.

How, without analogia and strange loops, would creativity be possible? Creativity is the generation of something new from something old; analogy sits at the very center of the process, as one note or word or brushstroke inspires the next, one idea births another, one old solution is modified to solve a new problem. All of these are analogical events; and all become, necessarily, strange loops, as there is no separating the created thing from the perceptions and experiences that inspired it – even if those perceptions or experiences were not consciously surfaced in the creative act.

"And that is also the way the human mind works," wrote Hofstadter in *Strange Loop*, "by the compounding of old ideas into new structures that become new ideas that can themselves be used in compounds, and round and round endlessly, growing ever more remote from the basic earthbound imagery that is each language's soil."

Standard components

Without *This* is like *That*, without strange loops, intelligence and consciousness could not exist. Learning and understanding, if possible, would be static and ineffective in our dynamic and ever-changing physical landscape.

We can say, then, that analogia and strange loops qualify as *inherent* properties of intelligence and consciousness. And if they are required in a biological mind, they likewise would be required in a machine mind; for the claims of the previous paragraph would still prevail.

An experiential strange loop engine, whether brain or box, becomes conscious, given enough energy, capacity, and experience. It becomes an *I*.

"When and only when such a loop arises in a brain or in any other substrate, is a person-a unique new 'I' - brought into being," Hofstadter wrote in *GEB*. "Moreover, the more self-referentially rich such a loop is, the more conscious is the self to which it gives rise. Yes, shocking though this might sound, consciousness is not an on/off phenomenon, but admits of degrees, grades, shades. Or, to put it more bluntly, there are bigger souls and smaller souls.

"This is a liberating shift, because it allows one to move to a different level of considering what brains are: as media that support complex patterns that mirror, albeit far from perfectly, the world, of which, needless to say, those brains are themselves denizens - and it is in the inevitable self-mirroring that arises, however impartial or imperfect it may be, that the strange loops of consciousness start to swirl."

Somewhere in Babel is a Doug Hofstadter who said *this* or *that* differently in this book or that one, this talk or that, and won the world over with analogia and strange loops as the correct model for how the mind works. And there would be billions of variations of those children of Babel reading that winning argument (or one of its endless variations), who would absorb its own strange loops and pass them on.

We can wonder if that Doug Hofstadter is any less melancholy...

Defining Requirements of Consciousness

Analogical Thought

Strange Loops

Douglas, Daniel, John: An Infernal Chinese Room

John Searle is a child of Babel.

The Berkeley philosopher is a galaxy-wide band of neuron-descriptive digits, articulating more than five decades of thoughts and writings and activity in academia and politics. He, like Borges and Hofstadter and HAL and all who are to follow, exists in both the actual and the almost, the box set of the real thing and billions of billions of near-misses. And for every one of these, there will be billions of billions of records of minds who heard his Chinese Room argument and agreed with it – or disavowed it. Much fewer will be those who had no opinion.

In this age of Siri, Alexa and their emerging sisters, and the endless stream of prompts, updates, and recommendations they provide us, we are beginning to take talking machines for granted. With the advent of "smart cloud" technology in the workplace, we will soon take digital assistants for granted - ethereal managers who keep us on schedule, screen our calls and emails, and review our work before it goes out.

But despite our casual acceptance of machine influence in our thoughts and actions, we realize that these novelties aren't actual thinking machines. Many decades after Alan Turing proposed his optimistic challenge, we have nothing out there (yet) on the digital landscape that is truly capable of passing for human.

We can, however, given our interactions with our digital helpers, see that the Turing Test (if you can't tell you're conversing with a machine, then the machine can be said to be 'thinking') may not be the true measure of the thinking machine.

Searle turned his attention to this question in 1980. On a flight to an academic conference, he created a thought experiment to demonstrate his conviction that no digital computer, however convincing in elocution and mannerism, could ever be said to truly think:

"Imagine that you are locked in a room, and in this room are several baskets full of Chinese symbols. Imagine that you (like me) do not understand a word of Chinese, but that you are given a rule book in English for manipulating these Chinese symbols. The rules specify the manipulations of the symbols purely formally, in terms of their syntax, not their semantics. So the rule might say: 'Take a squiggle-squiggle sign out of basket number one and put it next to a squiggle-squoggle sign from basket number two.' Now suppose that some other Chinese symbols are passed into the room, and that you are given further rules for passing back Chinese symbols out of the room. Suppose that unknown to you the symbols passed into the room are called 'questions' and the symbols you pass back out of the room are called 'answers to the questions'. Suppose, furthermore, that the programmers are so good at designing the programs and you are so good at manipulating the symbols, that very soon your answers are indistinguishable from those of a native Chinese speaker. There you are locked in your room shuffling your Chinese symbols and passing out Chinese symbols in response to incoming Chinese symbols... from the point of view of an outside observer, you behave exactly as if you understood Chinese, but all the same you don't understand a word of Chinese. But if going through the appropriate computer program for understanding Chinese is not enough to give *you* an understanding of Chinese, then it is not enough to give *any other digital computer* an understanding of Chinese."[13]

Put another way - if Searle, who has a human brain and all its linguistic and semantic potential, can functionally translate Chinese convincingly and *still* not understand Chinese, what chance does a microprocessor have?

Many were the outraged replies from the artificial intelligence community, when Searle put forth this argument:

The Systems Reply: "Neither the person in the room nor the rule book understand Chinese, but the *entire system* understands Chinese."

The Robot Reply: "The Chinese Room can be placed inside a robot, roam the landscape, and develop causal connections between the symbols and the objects they represent."

The Brain Simulator Reply: "Suppose the rule book is so finely detailed that it simulates the responses of every neuron in the brain of a

---

[13] *Minds, Brains and Science*, pp. 32-33.

native Chinese speaker. Then there is no distinction between the room and an actual Chinese mind."

And so on (there are many, many replies). The shorthand for the controversy boiled down to *The Chinese Room argument says that machines can never think/become conscious*, and this of course raised the hackles of countless academics, technologists and science fiction aficionados.

Strong AI and stuff

Hofstadter and Daniel Dennett of Tufts University included Searle's Chinese Room paper in their 1981 book *The Mind's I*, and offered their own two cents:

"Our response to this...is basically the 'Systems Reply': that it is a mistake to try to impute the understanding to the (incidentally) animate simulator; rather it belongs to the system as a whole, which includes what Searle casually characterizes as 'bits of paper.' This offhand comment, we feel, reveals how Searle's image has blinded him to the realities of the situation. A thinking computer is as repugnant to John Searle as non-Euclidean geometry was to its unwitting discoverer, Gerolamo Saccheri, who thoroughly disowned his creation."

The Chinese Room became the springboard for endless presentations of arguments in favor of artificial consciousness. Most these hinged upon convenient *ad hoc* definitions of the words *understanding*, *consciousness*, and *meaning*, and consequently there is very little order to the controversy: it is simultaneously one of the longest-running debates in academia and one of the most nonsensical.

Searle's actual position on machine consciousness:

"Another misunderstanding of the Chinese Room Argument is to suppose that I am arguing that as a matter of logic, as an *a priori* necessity, only brains can have consciousness and intentionality," he has written[4]. "But I make no such claim. The point is that we know in fact that brains do it causally. And from this it follows as a logical consequence that any other system that does it causally, i.e., that produces consciousness and intentionality, must have causal powers to do it at least equal to those of human and animal brains. But it does not follow that other systems have to have neurons to do it. (Compare: airplanes do not have to have feathers in order to fly, but they do share with birds the causal powers to overcome the force of gravity in the earth's atmosphere.) The question of which systems are causally capable of producing consciousness and intentionality is an empirical factual issue, not to be settled by *a priori* theorizing. Since we do not know exactly how brains do it we are in a poor position to figure out how other sorts of systems, natural or artificial might do it. *But there is no logical or metaphysical obstacle to consciousness and intentionality being caused in some other system, whether natural or artificial.*[14]"

Even so, if there is confusion on where Searle stands, it's nobody's fault but his: in spending so much time and energy talking about the Chinese Room, he has on many occasions resorted to exceedingly imprecise language (for a philosopher of language). Even in his responses to the replies published alongside the original paper[15], he wrote things like:

"Given the coherence of the animal's behavior and the assumption of the same causal stuff underlying it, we assume both that the animal must have mental states underlying its behavior, and that the mental states must be produced by mechanisms made out of the stuff that is like our stuff."

Stuff?

This stuck in Hofstadter's craw, who wrote (a full 17 years later[16]), "...Searle, in his insistence on the genuine semanticity of human language, is an inveterate, dyed-in-the-wool believer in the idea that words that come tripping off human tongues *really refer to the world*, but not words that trip across the screen of a transistorized machine.

"And this ineluctable black-and-white dichotomy, we are told, comes about thanks to mysterious 'causal powers of the brain' not shared by computers guided by programs. How does Searle know? Oh, he just *does*. As he puts it, computers are made of 'the wrong stuff'. 'Wrong stuff', 'right stuff', 'causal powers of the brain' - it's all just too surreal for me."

Hofstadter's impatience with linguistic imprecision is not confined to Searle, but proffered to philosophers in general: they are "players with words," he told John Horgan. Fair enough, but it becomes noise obscuring our signal; in our quest to pin down inherent properties of consciousness, it's an *ad hominem* distraction that tugs us down a rabbit hole.

Another rabbit hole frequently traversed by those arguing over the Chinese Room is that its primary intent is to disprove "Strong AI", which is customarily defined as the view that "the appropriately programmed

---

[14] In the *Blackwell Companion to the Philosophy of Mind*, 1994.

[15] Searle 1980.

[16] In *Le Ton Beau de Marot: In Praise of the Beauty of Language*, 1997.

digital computer with the right inputs and outputs would thereby have a mind in exactly the same sense that human beings have minds." Those are Searle's words, from his book *Mind, Language and Society*; he, in fact, coined the term, which is now in general use.

Defenders of Strong AI (Hofstadter and Dennett[17] among them) tend to leap at Searle's perceived attack and reduce the Chinese Room argument to the defense of Strong AI alone; but Searle (and his own defenders) perpetually point out that 1) "Strong AI", in his formulation, only applies to digital computers as we experience them today, i.e. pure symbol processors performing purely algorithmic operations; 2) the Chinese Room is about what's going on (or not) *inside* the room, not what the outside observer experiences; and 3) the point is to clarify what it means to *understand* – without which, all speculations aside, no serious pursuit of artificial consciousness will be possible in the first place.

Put another way, the refutation of Strong AI by the Chinese Room argument is a peripheral concern at best; the primary intent is to underscore what is most essential in achieving AI that truly rises to the level of human thought.

A conscious Chinese Room

We must also clarify that the Chinese Room, which at face value is about "understanding", is ultimately addressing the nature of

---

[17] Dennett, like Hofstadter, still rails against the Chinese Room, decades on: in his excellent 2013 *Intuition Pumps and Other Tools for Thinking*, he calls out the Chinese Room as a faulty exemplar of an *intuition pump* (simple thought experiments that inspire contemplation about their theses), calling it "a boom crutch that can disable your imagination..." He front-loads his argument with the usual misquotes of Searle's intent for his argument, then drifts into discourse, not about machine consciousness, but about machine *competence*: "The way to reproduce human competence and hence comprehension (eventually) is to stack virtual machines on top of virtual machines on top of virtual machines – the power is in the system, not in the underlying hardware." If competence somehow inherently leads to comprehension, Dennett makes no effort to articulate how this might be; and if competence is truly the path to machine consciousness, we should already be there, for machine competence has already greatly surpassed human competence in dozens of domains far beyond chess and Go. And, once again, Dennett takes not a single step toward addressing Searle's claim that semantics, not syntax, lies at the heart of consciousness. [As a point of interest, Dennett actually coined the term *intuition pump* in his original reply to the Chinese Room, and has applied it to dozens of thought experiments since.]

consciousness, which Searle's original paper does not mention (though it pops up repeatedly in the first generation of replies).

Searle has long since dispelled all doubt.

"The commonsense objection to strong AI was simply that the computational model of the mind left out the crucial things about the mind such as consciousness and intentionality," he later wrote[18]. "Consciousness is the central fact of specifically human existence because without it all of the other specifically human aspects of our existence – language, love, humor, and so on – would be impossible."[19]

Philosopher David Chalmers has likewise recognized the intent of Searle's argument: "...Searle directs the argument against machine *intentionality* rather than machine consciousness, arguing that it is 'understanding' that the Chinese room lacks. All the same, it is fairly clear that consciousness is at the root of the matter. What the core of the argument establishes directly, if it succeeds, is that the Chinese room system lacks conscious states, such as the conscious experience of understanding Chinese. On Searle's view, intentionality requires consciousness, so this is enough to see that the room lacks intentionality also."[20]

Intentionality now makes its way into the mix, and we can add it to our list of components we must consider when contemplating the possibility of conscious AI. Pierre Jacob defines it as "the power of minds to be about, to represent, or to stand for, things, properties and states of affairs"[21]. Searle defines *understanding* in the original Chinese Room argument as *mental states with intentionality*, and those are artifacts of consciousness.

Searle himself paraphrases Jacob as follows: "...that property of many mental states and events by which they are directed at or about or of objects and states of the world.

"If, for example, I have a belief, it must be a belief that such and such is the case; if I have a fear, it must be a fear of something or that something will occur; if I have a desire, it must be a desire to do something or that something should happen or be the case; if I have an

---

[18] In *The Rediscovery of the Mind*, 1992.

[19] In *Minds, Brains and Science*, 1984.

[20] In *The Conscious Mind*, 1996. Chalmers was once Hofstadter's student, and the two of them parted company over their fundamental disagreement on the nature of consciousness.

[21] In the *Stanford Encyclopedia of Philosophy*.

intention, it must be an intention, it must be an intention to do something."[22]

Dan Dennett returns to the conversation at this point with his own variation, the *intentional stance*, which takes Searle's more classical belief and operationalizes it in interesting ways:

"Here is how it works," he wrote. "First you decide to treat the object whose behavior is to be predicted as a rational agent; then you figure out what beliefs that agent ought to have, given its place in the world and its purpose. Then you figure out what desires it ought to have, on the same considerations, and finally you predict that this rational agent will act to further its goals in the light of its beliefs. A little practical reasoning from the chosen set of beliefs and desires will in most instances yield a decision about what the agent ought to do; that is what you predict the agent will do."[23]

In both definitions, the individual is associating beliefs with objects, events and actions; in Dennett's case, these associations take on a predictive value when applied to others. And for both Dennett and Searle, intentionality is inextricably bound to consciousness[24].

Having established that the Chinese Room is in fact a conversation about consciousness (which, we will stipulate, must be present to enable intentionality), we must now zero in on a strong working definition of "consciousness" - one that accurately describes human beings and has measurable utility in a machine implementation.

Stuart Sutherland offers the following, from the *Macmillan Dictionary of Psychology*: "Consciousness - The having of perceptions, thoughts, and

---

[22] In *Intentionality: An Essay in the Philosophy of Mind*, 1983.

[23] In *The Intentional Stance*, 1987.

[24] As an interesting aside, Dennett and Searle briefly became strange bedfellows in the book *Neuroscience & Philosophy*, in which they defend the enterprise of cognitive neuroscience against an assault by neuroscientist Maxwell Bennett and philosopher Peter Hacker. The latter two argue that experimental inquiry into what features of consciousness brains enable is pointless and doomed, as coherent questions about them cannot be framed in the language of science (conveniently, they believe that only philosophy can suffice as a framing language). Dennett and Searle, of course, disagree completely, maintaining that the scientific inquiry into consciousness is of incalculable benefit. At stake is whether the integration of the study of human nature is essential to the success of cognitive neuroscience, and if so, whether it is possible.

feelings; awareness. The term is impossible to define except in terms that are unintelligible without a grasp of what consciousness means. Many fall into the trap of equating consciousness with self-consciousness - to be conscious it is only necessary to be aware of the external world. Consciousness is a fascinating but elusive phenomenon: it is impossible to specify what it is, what it does, or why it has evolved. Nothing worth reading has been written on it."

More useful, perhaps, is the definition offered by the *Routledge Encyclopedia of Philosophy*: "Philosophers have used the term 'consciousness' for four main topics: knowledge in general, intentionality, introspection (and the knowledge it specifically generates) and phenomenal experience... Something within one's mind is 'introspectively conscious' just in case one introspects it (or is poised to do so). Introspection is often thought to deliver one's primary knowledge of one's mental life. An experience or other mental entity is 'phenomenally conscious' just in case there is 'something it is like' for one to have it. The clearest examples are: perceptual experience, such as tastings and seeings; bodily-sensational experiences, such as those of pains, tickles and itches; imaginative experiences, such as those of one's own actions or perceptions; and streams of thought, as in the experience of thinking 'in words' or 'in images'. Introspection and phenomenality seem independent, or dissociable, although this is controversial."

For our purposes – considering machines that might qualify as conscious in the sense that we ourselves are – we might pare these down to "self-awareness in the form of an internal observer that can monitor its own experience, comparing it to past experience and the experience of others, in order to anticipate future experience." This limited definition focuses on utility, which is what we will be seeking in the quest for conscious AI.

The meaning of 'meaning'

We're left with this: for all the noise about Strong AI and "stuff" and faulty intuition pumps, Searle was trying to make a single point above all others with the Chinese Room argument, and no reply has ever met it head-on: *syntax is insufficient for consciousness; semantics are required.*

"Because the program is purely formal or syntactical and because minds have mental or semantic contents," he concluded, "any attempt to produce a mind purely with computer programs leaves out the essential features of the mind."

"Computational models of consciousness are not sufficient by themselves for consciousness," he wrote. "The computational model for consciousness stands to consciousness in the same way the computational model of anything stands to the domain being modeled. Nobody supposes that the computational model of rainstorms in London will leave us all wet. But they make the mistake of supposing that the computational model of consciousness is somehow conscious. It is the same mistake in both cases."[25]

Searle's opponents are recycling the assumption of Turing – that the output of a conscious system is evidence of its consciousness, an assumption which does not take into account the possibility that a system that is not conscious could produce output indistinguishable from that of a conscious one. That's a hard assumption to defend: in the year of this writing, the Internet is populated by millions of chatbots whose communication skills are more than enough to pass for human in the minds of tens of millions of consumers – and which are no more 'conscious' or 'thinking' than an Excel spreadsheet. It's an assumption rooted in the Theory of Mind – the attribution of mental states (beliefs, emotions, intentions, knowledge) to ourselves and others like ourselves – such that when we are on the receiving end of linguistic expressions that resemble our own, we instinctively attribute mental states to the source of those expressions (we must not fail to realize that this is a near-perfect example of Hofstadter's *analogia*.)

Let's take Searle at face value, and limit our inquiry to the role of semantics in consciousness, regardless of platform.

Let's start with the chatbot.

Expressing itself purely syntactically, according to a programmed script with many conversational branches, a chatbot is able to convincingly mimic a human being. While tens of millions of Internet-surfing consumers are now aware of chatbots and even assume them to be so when they pop up, there was a learning curve there – and most of us, at some point, transitioned to knowing a chatbot when we saw one from assuming the pop-up was a human being. Today, millions are still unaware they are chatting with a program, so effective is their performance.

The meaning in a conversation between a human being and a chatbot resides in the mind of the human; there is none in the chatbot itself, nor

---

[25] In *Consciousness and Language*, p. 16.

would we expect to find any, as chatbots are designed and deployed not to be conscious or to have a meaningful existence, but to extract information from human users. When we describe the chatbot as purely syntactical, we can feel confident we're on safe ground. The words received by the chatbot from the human and returned to the human from the chatbot are pure symbols, mapped together in the same conceptual fashion as the Chinese Room rules. Put another way, the content of the exchange – words – have semantic value to the human, but not to the chatbot.

But wait – can we be certain? Isn't a 'meaning' a definition? *Of course* the chatbot has access to definitions. A customer service chatbot knows the contextual parameters of *service*, and necessarily knows the uses and features of products, or it couldn't effectively respond. Doesn't that qualify as *meaning*?

We can say the same of a dictionary. Every word in a dictionary is defined by other words in the dictionary, giving it endless circularity. The dictionary is nothing *but* definitions – but absent a reader, a dictionary is *devoid* of meaning. When we open a dictionary and read the definition of a word, we get the meaning when we map that definition to our experience of the object or idea being referenced – or, if it is a new word, doing a mapping to some other object or idea we have experienced (in this sense, dictionaries are yet another near-perfect expression of Hofstadter's analogia). We can say the chatbot cannot create meaning from definition because chatbots don't experience anything.

Or do they? When a chatbot has an exchange with a user, the exchange is typically recorded – stored as data for future reference. Higher-quality chatbots are able to reference previous exchanges with a particular user when that user later re-engages. Doesn't that qualify as *experience*?

Unfortunately, no; a stored exchange with a user only includes the chatbot's experience with the user; it does not represent experience with anything of substance in the exchange. It has never received services as a customer, or used any of the company's products. It can only reference those things as a dictionary refers to its own words. And those stored exchanges, when retrieved, serve only as inputs into the bot's script-parsing process; they do not alter its programming in any way, as the neural activity caused by new experiences alters own own.

But wait – many chatbots now have access to AI engines, and are thus able to learn. Specifically, AI-based bots can interpret the user's emotional state; they are able to tell, for instance, if a customer is satisfied or irate. This interpretation represents *meaning*, doesn't it? The bot is able to identify the sentiment beneath the customer's words, and respond accordingly. Moreover, the bot gets better at this over time – isn't that *experience*?

Again, unfortunately, no; while the perceived emotions of others certainly have meaning to us humans, it can't be the case with a bot. Why? Because in the bot, the learned sentiment interpretation is a *definition*, not a *meaning*; it can experience a user's emotional expression, but it *cannot experience that emotion*. A bot has no mental states; it cannot be satisfied or irate, or frustrated or confused or amused or anything else a user can be. When we interpret the sentiment of another, a skill we likewise improve over time, the meaning of the sentiment is clear to us because it maps to our own experienced sentiments. That's not the case with a chatbot.

In short, the chatbot does not operate from the Theory of Mind. No matter how sophisticated its responses, no matter how skillful its emotional interpretation, no matter how effective its performance, it is not experiencing, and so it is not assigning meaning. It is merely a Chinese Room of another sort.

An English Room, a Searle AI

Searle adds another interesting wrinkle to his assertion about the Chinese Room. He reframes the problem as follows: he leaves the Chinese Room and goes into an English Room. English, of course, he understands perfectly well; he asks the reader to simply compare the two scenarios side-by-side. What's the difference between the Chinese Room and the English Room? That's the difference that defines consciousness, he claims. A truly conscious AI is analogous to the English Room, but not the Chinese Room.

Where can this take us? Let's play it out: let's say that Searle's original Chinese Room is indeed an analogy of a digital computer – and that the English Room is an analogy of a conscious AI. Now let's take the next step: let's replace the Chinese Room Searle with an actual conscious AI that does not understand Chinese.

To say that the Searle AI doesn't understand Chinese is to say that it is unable to map the syntax of the inbound messages and the rule book to its own conscious experience or understanding. Despite being conscious, it would be incapable of creating analogies – there would be no *this* is like *that*. It can have no intentionality, regarding the contents of the inbound messages or the rules and how they impact its beliefs or mental states, as it knows nothing of their semantic substance.

In this scenario, the Systems Reply collapses; for even though the room contains a conscious component, and even though the component is an AI, *the system itself is* not *conscious*: it is, as a system, absent of semantics, absent of meaning, absent of intentionality - and such experience as the conscious AI might possess is disparate from the room itself.

No strange loops

The Searle AI version of the Chinese Room cannot bring analogy into the mix, and likewise cannot experience strange loops.

If strange loops are a defining component of consciousness, and if the Searle AI is conscious, then the Searle AI possesses strange loops, by definition. That is, it has exchanged bits and pieces of its knowledge, insights, and experiences with other conscious entities, be they human or not, and has built its consciousness from those loops; we can say, then, that a component of this Chinese Room possesses strange loops.

But the room itself – no. Sealed within the Chinese Room, the Searle AI *has no other conscious entity with which to exchange its knowledge, insights, and experiences*; no new strange loops, then, are forming *within the system*, because what is being shared by agents outside the room – the inbound symbols – do not represent knowledge, insight, or experience.

Here's an interesting twist: that outside agent, interacting with the Chinese Room, *is* able to potentially derive new strange loops from the interaction. We presume the outside agent to be both conscious and a speaker of Chinese, interacting with the room in order to receive answers to questions.

Presuming the accuracy of the rule book (that the answers to the questions are meaningful to a Chinese speaker), we can see that it is possible for such answers to have meaning to the person outside the room (who could well be a conscious AI) - perhaps profound meaning, and thus the author of the rule book may pass strange loops to the outside agent. It is necessarily a one-way road; no strange loops within the outside agent have an inbound path, for no matter how rich and meaningful its questions, neither the room nor its conscious AI are able to distinguish the outside agent's questions from noise.

And that's before we consider that any strange loop uptake within the room would need to be reflected in the rule book, in order for the *system* to be considered conscious.

In the English Room variation, the English-speaking John Searle may receive inbound questions and be heavily affected by their potential richness and meaning; he might be moved to ponder the meaning of life in new ways, to take up new and innovative lines of inquiry, or realize the errors of his youth anew, based on the content of the questions. This qualifies him as fully conscious – but according to the specification of the room, none of this has any bearing on the room's output. Searle understanding English and forming new strange loops doesn't advance the consciousness of the system in any way – it is still producing static, predetermined responses, per the rule book.

A final word or two on the Systems Reply

We've been pretty hard on the Systems Reply – but then, its advocates have been pretty hard on Searle. The persistent claim that entire systems comprised of static components are conscious is unfathomable when we extend the concept beyond the Chinese Room; it's like believing that corporations truly are people.

Even so, let's take one final step into the Room. Let's imagine that the active agent in the room is neither Searle nor an AI, but a digital robot – and that the *rule book* is a conscious AI.

We are still bound by the rules of the Room: regardless of the rule book being conscious (and speaking Chinese), its *function* remains the static, algorithmic mapping of symbols. The digital robot, the Room's only moving part, is just as static and algorithmic. The conscious Rule Book AI is self-aware, experiencing, possibly thinking up analogies and forming intentional beliefs and desires all day long; but the only thing it is doing, as a component of the system, is mapping symbols per the rules it knows. The conditions of the Room are the same as they've always been.

The Rule Book AI is conscious; is the *system* conscious?

The Room, the system, is still not self-aware. The system is not experiencing anything. It is not thinking up analogies or forming intentional beliefs or desires. It is still performing algorithmically, statically, offering the same output y to input x. It is not learning, it cannot improvise, despite the consciousness that hosts its algorithms (which is capable of these things, but constrained by the algorithmic nature of the Room).

Now let's go all-in: replace the digital robot with the Searle AI, leaving the Rule Book AI in place.

In doing so, we fulfill another requirement of consciousness: we've put two conscious entities in the Room, with the supposition that they will interact. Earlier, we've established that consciousness dawns and flourishes when conscious individuals interact, sharing their cognitions, *exchanging strange loops!* – and we've introduced that feature into the Room. (For that matter, we could replace the digital robot with Searle himself, of course – or make Searle the rule book, having memorized the mappings.)

Even with two conscious beings in the Room, the Room itself – the system! - is *still* not self-aware. The system is *still* not experiencing anything. It is *still* not thinking up analogies or forming intentional beliefs or desires. It is *still* performing algorithmically, statically, offering the same output y to input x. It is *still* not learning, it *still* cannot improvise, despite the two conscious components doing *all* the work.

Today, I can do something Searle and Hofstadter and Dennett couldn't do in 1980; I can actually test the Systems Reply right now, sitting at my desk.

I can create a Chinese Room on demand. I can pick up my iPhone and invoke Siri; she is, after all, a digital AI with access to a rule book. I have a CD of Mandarin tourist questions; Siri speaks Mandarin (I still don't). I can repeat the tourist questions to Siri and receive answers in Mandarin.

I, of course, am conscious. Siri is not; the CD is not. Is the *system* formed by the three of us conscious? Would anyone imagine that the system we form has self-awareness independent of mine, intentionality beyond mine? Are there now *two* consciousnesses present, here at my desk?

I ask my questions of Siri. Siri answers in Mandarin. But nowhere in this process has any conscious understanding of Mandarin occurred whatsoever, within the system as a whole or any of its components.

The Chinese Robot

In Searle's original paper, republished in *The Mind's I*, we find several replies to his argument that are (unlike the Systems Reply) conciliatory. The Robot Reply, for instance, concedes that symbol manipulation alone is insufficient for consciousness – but that the Chinese Room would be, if it were implemented as the brain of an ambulatory robot:

"Suppose we put a computer inside a robot, and this computer would not just take in formal symbols as input and give out formal symbols as output, but rather would actually operate the robot in such a way that the robot does something very much like perceiving, walking, moving about, hammering nails, eating, drinking – anything you like. The robot would, for example, have a television camera attached to it that enabled it to see, it would have arms and legs that enabled it to 'act,' and all of this would be controlled by its computer 'brain.' Such a robot would...have genuine understanding and other mental states."[26]

Searle's answer: "Suppose that instead of the computer inside the robot, you put me inside the room and, as in the original Chinese case, you give me more Chinese symbols with more instructions in English for matching Chinese symbols to Chinese symbols and feeding back Chinese symbols to the outside. Suppose, unknown to me, some of the Chinese symbols that come to me come from a television camera attached to the robot and other Chinese symbols that I am giving out serve to make the motors inside the robot move the robot's legs or arms. It is important to emphasize that all I am doing is manipulating formal symbols: I know none of these other facts. I am receiving 'information' from the robot's 'perceptual' apparatus, and I am giving out 'instructions' to its motor apparatus without knowing either of these facts. I am the robot's homunculus, but unlike the traditional homunculus, I don't know what's going on., I don't understand anything except the rules for symbol manipulation. Now in this case I want to say that the robot has no intentional states at all; it is simply moving about as a result of its electrical wiring and its program. And furthermore, by instantiating the program I have no intentional states of the relevant type. All I do is follow formal instructions about manipulating formal symbols."[27]

Now we're getting somewhere: Searle is keeping the exchange on track by emphasizing that the introduction of causal associations still doesn't introduce intentionality or self-awareness, despite the fact that the robot is able to generate experience – but at least experience is now on the table.

We have also made a place for algorithms, after a fashion, in conscious systems: Searle's assessment of how inputs and outputs facilitate the mobility required for the system to navigate a landscape and generate causal associations through experience is spot-on (though not in the way he intends) in that it need not impose on a conscious system's awareness; for the most part, we give no conscious thought to the inputs and outputs traversing our brains when we keep our balance while rising to our feet

---

[26] *The Mind's I*, p. 362.

[27] Ibid., pp. 362-3.

from a chair, reaching for and turning a doorknob, and so on; these acts are generally reflexive, and for the most part, require little or no self-awareness.

Even so, a role for algorithms in a conscious system doesn't set aside Searle's point, which is that semantics are still missing from the mix; but he unfairly fails to acknowledge that the Robot Reply is extending a conciliatory olive branch on this point. *The robot has the means to add meaning to the Chinese symbols by experiencing the world they describe*; the reply simply doesn't take the necessary step of exploring the possibilities for integrating that meaning into the system. Absent such a mechanism, Searle's objections hold firm.

The Robot Reply doesn't rebut the Chinese Room; but it does introduce a new potential inherent feature of consciousness: *Mobility*, which we'll examine in the next chapter.

A Chinese Brain

Then there's the Brain Simulator Reply, likewise reprinted in *The Mind's I*, which takes a bold step away from the tenets of Strong AI:

"Suppose we design a program that doesn't represent information that we have about the world.. But simulates the actual sequence of neuron firings at the synapses of the brain of a native Chinese speaker when he understands stories in Chinese and givers answers to them. The machine takes in Chinese stories and questions about them as input, it simulates the formal structure of actual Chinese brains in processing these stories, and it gives out Chinese answers as outputs. We can even imagine that the machine operates, not with a single serial program, but with a whole set of programs operating in parallel, in the manner that actual human brains presumably operate when they process natural language. Now surely in such a case we would have to say that the machine understood the stories; and if we refuse to say that, wouldn't we also have to deny that native Chinese speakers understood the stories? At the level of the synapses, what would or could be different about the program of the computer and the program of the Chinese brain?"[28]

Before we dig into that, here's Searle:

"...getting this close to the operation of the brain is still not sufficient to produce understanding. To see this, imagine that instead of a

---

[28] Ibid., p. 363.

monolingual man in a room shuffling symbols we have a man operate an elaborate set of water pipes with valves connecting them. When the man receives the Chinese symbols, he looks up in the program, written in English, which valves he has to turn on and off. Each water connection corresponds to a synapse in the Chinese brain, and the whole system is rigged up so that after doing all the right firings, that is after turning on all the right faucets, the Chinese answers pop out at the output end of the series of pipes.

"Now where is the understanding in this system? It takes Chinese as input, it simulates the formal structure of the synapses of the Chinese brain, and it gives Chinese as output. But the man certainly doesn't understand Chinese, and neither do the water pipes, and if we are tempted to adopt what I think is the absurd view that somehow the conjunction of man and water pipes understands, remember that in principle the man can internalize the formal structure of the water pipes and do all the 'neuron firings' in his imagination. The problem with the brain simulator is that it is simulating the wrong things about the brain. As long as it simulates only the formal structure of the sequence of neuron firings at the synapses, it won't have simulated what matters about the brain, namely its causal properties, its ability to produce intentional states. And that the formal properties are not sufficient for the causal properties is shown by the water pipe example: we can have all the formal properties carved off from the relevant neurobiological causal properties."[29]

Searle prefaces his rebuttal of the Brain Simulator Reply with a note of astonishment that it was even offered, given that it breaks with Strong AI altogether in abandoning straightforward algorithmic processing – conventional digital programming – in favor of what we today call *neural networks*[30]. But the rebuttal itself puts him on his shakiest ground yet, because he misses the point of the reply: the proposed neural network does not overlay the English rules the monolinguist in the room is using; *it replaces both the rules and the person.*

---

[29] Ibid., pp. 363-364.

[30] Neural networks will be discussed in depth in the following chapter, but in a nutshell, they are *connectionist systems* – layers of interconnected nodes that send signals forward in the manner of neurons and synapses, with weights embedded in the nodes that adjust signal strength over time, refining the outputs as the system learns. While neural networks are indeed ultimately algorithmic when simulated on a digital computer, an actual neural network physically realized in a microprocessor would bear very little resemblance to a computer: it would learn and adapt and perform much more like a human brain.

In fairness, at the time of the reply (1980), neural networks weren't yet a thing: they were just intellectual exercises, written off largely thanks to Marvin Minsky, who demonstrated in the late Sixties that they could not solve Exclusive Or problems. This changed in 1986, when Rumelhart and McClelland demonstrated that Minsky's limitation only applied to two-layer networks. Since then, neural network technology has flourished, and is the engine of what we call *deep learning* today – our highest, most powerful level of AI.[31]

In such a system, there is no "program" in the sense that Searle has been using the word – and the information processing mechanism being used to respond to the incoming Chinese symbols is not static; it changes with every interaction, constantly learning and improving. This being the case, Searle's objections regarding the cluelessness of the man in the room must necessarily vanish, as the man himself vanishes.

The pipe-and-valve system Searle describes in his rebuttal is a fair (if simplistic) analogy for the activity of neurons and synapses, but even so, it completely misses the point when it calls for a man to turn valves on and off per instructions from a rule book: in a neural network, there *is* no rule book.

Moreover, if the neural network described is in fact an instantiation of the brain of a native Chinese speaker, then it necessarily has captured the *meanings* of the symbols being translated, not just the syntax; for a native Chinese speaker cannot learn to speak Chinese without making those semantic connections.

In fairness to Searle, the system described in the Brain Simulator Reply was insufficiently articulated, and the inner workings of neural networks were unknown even to most computer experts in 1980. Their potential was not even remotely anticipated.

Even so, Searle is perfectly correct, although inadvertently; there is a difference-in-kind between a *simulated* neural network and an *actual* neural network, and it informs our pursuit of the inherent qualities of conscious AI.

A *simulated* neural network is, in fact, a digital program of exactly the sort Searle dismisses – a very lengthy series of computer instructions that mimics the activity of nodal layers that instantiate the network. The results, of course, are the same, input- and output-wise, as one would get

---

[31] *Deep learning* is performed by complex multi-layer neural networks, and is used in hundreds of real-world applications, from image processing to voice and facial recognition to natural language processing.

from an actual network; but all of Searle's objections to algorithmic processing come flooding back if we don't make this distinction.

In an *actual* neural network, there is no mimicry; there is a circuit containing layered nodes that receive and pass along signals through hardwired connections. They are literally analogs to human neurons and synapses, wrought in copper and silicon. *None* of Searle's objections can apply to this architecture, as there are no commands, no rules – *no syntax*.

Neural networks are an embodiment of a concept that Searle skips right over: *distributed processing*, or many things happening at once. In every variation of the Chinese Room considered so far, the active agent in the Room executing the translation rules is doing so serially, one at a time; that's a deal-breaking constraint that kills the prospect of consciousness. If Searle had possessed a more complete understanding of the distributed processing aspect of proposed network, he could have leveraged that as a contrast that strengthened his rebuttal.

Finally, the authors of the reply bring in yet another useful component: *parallel processing*, over and above *distributed processing*:

"We can even imagine that the machine operates, not with a single serial program, but with a whole set of programs operating in parallel, in the manner that actual human brains presumably operate when they process natural language."

Distributed processing is what an individual network does, spreading the work over many nodes; *parallel* processing is several networks all operating together, with connections between them, using their outputs as inputs to other networks and accepting the outputs of other networks as their inputs, all in real time.

Both of these describe what happens in a conscious human brain; both can be implemented as technology. Both can now take their place on our list of inherent properties of consciousness – and we'll take them up in the next chapter.

For now, however, when we consider that almost every neural network that exists is a simulation, we must grant that Searle's bulwark remains unbreeched.

The essentials

The Chinese Room as originally presented, lacks semantics/
meaning; intentionality; experience; the ability to form analogies; strange loops. That's a pretty long list. And finally, there's Searle's last word on

the subject: the Chinese Room, as a system, *isn't aware it's speaking Chinese*. The absence of self-awareness underlies the absence of all the others.

It does have one essential feature of consciousness: intelligence. Its responses to inputs affect its outputs. That, ironically, is a feature it shares with - wait for it! - digital computers. We can work backwards from this, and produce an interesting corollary: while semantics are necessary for consciousness, they are *not* necessary for intelligence; *intelligence can be purely syntactical*.

If consciousness requires semantics, intentionality, experience, analogia and strange loops, and a digital computer can't provide those things, we are left with Searle having dangled the possibility of a non-biological platform for consciousness based on something else.

So, what kind of machine *could* host the ingredients of consciousness?

Based on what we've reviewed so far, such a machine would need to have experiences; it would need mechanisms by which to perceive its environment and interact with it in real time; more than that, it would need the opportunity to interact with other agents in the world. It would need perpetual access to its past and present, to retrieve its experiences and plan its future. It would need to process its experiences non-algorithmically, as neural networks do, such that all new experiences change its processes. It would need a mechanism by which it could self-monitor its experience of having experiences – awareness – and that, too, would need to be a non-algorithmic learning system. It would need some mechanism by which it could infer such awareness in other conscious agents, and discern the contents of their awareness.

Is such a machine possible? In principle, yes. But it's *way* more than a Chinese Room.

And there's one final feature of the conscious mind that any machine would require, one that hasn't been addressed.

The Chinese Room, like a digital computer, turns on and off. Conscious beings don't; brains run perpetually, from birth to death. Even at sleep, our minds are still working, though our senses go passive; even in a coma, our brains are still at work, keeping the body going. The perpetual nature of real neural systems, with their many layers, is wedded to the continuity of their experience and the mechanisms that make strange loops possible. Take that away, and the integrity of the connections that integrate consciousness might be at risk. Remember that HAL 9000's greatest terror, the one that humanized him the most and was ultimately his undoing, was being turned off.

We'll take this up later, downstream; moving forward, we have other fish to fry.

For the moment, we've identified some significant inherent properties of consciousness, and considered them in relation to the machine. In pondering the Chinese Room, we have discovered that it shines far more light than previously might have been supposed, and are left with one obvious but poignant conclusion: conscious or not, the Chinese Room is a child of Babel.

Defining Requirements of Consciousness

Analogical Thought

Strange Loops

Meaning

Experience

Intentionality

Distributed Processing

Parallel Processing

To HAL and Back

HAL 9000 is a child of Babel.

He (and his many variations) are unlike any other children of Babel, in interesting and insight-inspiring ways. But before we dive into that, let's quickly revisit the events of *2001: A Space Odyssey*...

In 1992 or 1997 (depending on whether you're reading the book or watching the movie), HAL 9000 Production Number 3 became operational at the HAL plant in Urbana, Illinois. Hal is highly intelligent, a supreme multitasker, and self-aware - able to not only converse with human beings, but to discuss problems and cooperate in solving them, as if he himself were human with digital senses and appendages. He is installed in the spaceship *Discovery*, being assembled in Earth orbit for a mission to Jupiter (or Saturn, again depending on whether you're consulting the movie or the book).

The idea behind his inclusion on the mission is to have technology in place that could conceivably complete the mission and return *Discovery* to Earth in the event of a catastrophe that killed or incapacitated the crew. That makes sense on any number of levels - but even more so when the nature of *Discovery*'s mission changed in 1999, when a mysterious black monolith was discovered in the crater Tycho on the moon.

Hal is informed of the true nature of the mission. The monolith had beamed a powerful signal to Jupiter, where *Discovery* was already headed when the monolith was found. The mission is now about possible first contact with an intelligence far more powerful than humanity. The ship's human crew - three scientists and two astronauts - were effectively placed in Hal 's keeping, as he and the three scientists (who were placed in hibernation for the journey) knew about the monolith and the astronauts did not (they were kept in the dark, for fear that thinking about such a thing for months in isolated space might compromise their sanity).

En route to Jupiter, Hal has developed an amiable working relationship with astronauts Bowman and Poole, both PhDs in their own right and intellectually compatible with the sentient computer. In an interview with an Earthbound journalist, Hal speaks as a human might, expressing collegial approval of Bowman and Poole and even suggesting a kind of pride in his own contribution to the mission.

We learn that Hal oversees every aspect of the ship's operation, almost rendering Bowman and Poole redundant. He can play chess very well, and is able to understand and even render judgments of Bowman's amateur artistic efforts.

Hal is an altogether triumphant exemplar of artificial intelligence - supremely competent, self-aware, and human-friendly. But a dark layer is building beneath his service: the mission planners on Earth have ordered him to lie to Bowman and Poole - to keep from them the true nature of their mission - and this causes Hal to lose his grip on reality. He makes a mistake - a faulty prediction.

When Bowman and Poole realize that Hal is malfunctioning, they quietly conspire to disconnect his higher cognitive functions - to effectively lobotomize him - and he learns these intentions by reading their lips. Concluding that Bowman and Poole now represent a threat to the mission's success, he proceeds to kill Poole during an extravehicular event, and attempts to do the same when Bowman leaves the ship to retrieve Poole.

Bowman perseveres, however, re-entering the ship in a spacesuit and proceeding to the room where Hal's cognitive circuitry resides. Hal attempts to reason with Bowman, growing increasingly desperate - even, by his own admission, afraid - as his disconnection looms. Bowman pulls drive after drive out of Hal's memory, as the computer's awareness begins to fail, and he slips into infantile memories as Bowman coolly pacifies him. Moments later, Hal L is dead.

A decade later, a joint US-Russian mission returns to *Discovery*, left in a parking orbit around Io, and Hal's original programmer - Dr. Chandra - awakens him after removing the memories of his homicidal behavior and the deaths of his crewmates. Chandra restores Hal to cognitive and emotional health, and Hal is instrumental in helping the mission avert disaster when the monolith ignites Jupiter, converting it into a mini-sun. Hal is liberated from his digital body by the monolith and is reunited with Bowman, who has evolved beyond human form - also facilitated by the monolith.

We can say with considerable confidence that Hal offers us much to work with in pondering the essentials of machine consciousness. The Hal narrative, written by Arthur C. Clarke in collaboration with venerated filmmaker Stanley Kubrick, may have been written half a
century ago, when AI was still a gleam in the eye and computers were still positively paleolithic, but there are nonetheless a great many intuitive contributions his story can make to our real-world assessment of the prospect of conscious AI.

Two quick exercises can get us rolling: the first is to profile HAL as a member of *Discovery*'s crew, to get a feel for just how conscious he is; the second is to list the ingredients of his consciousness that we find in the Clarke book and the Kubrick film.

Hal pals

Hal's participation in the *Discovery* mission, on the one hand, includes oversight of all the ship's operations (the "brain and central nervous system of the ship," per journalist Martin Amer), monitoring the men in hibernation, accommodating the requests of Bowman and Poole, providing them with recreation (the occasional game of chess), even tracking their psychological well-being (working up regular "crew psychology reports") - and rendering birthday greetings.

From Martin Amer's interview:

Amer: "In talking to the computer, one gets the sense that he is capable of emotional responses. For example, when I asked him about his abilities, I sensed a certain pride in his answer about his accuracy and perfection. Do you believe that Hal has genuine emotions?"

Poole: "Well, he acts like he has genuine emotions. Of course he's programmed that way to make it easier for us to talk to him. But as to whether or not he has real feelings is something I don't think anyone can truthfully answer."

In his interactions with Frank and Dave, Hal seems as human as Dr. Hunter, Kimball, or Kaminski would seem if they were having the same conversations. In a game of chess with Poole:

"Yeah," the latter replies to Hal's mate analysis, "Looks like you're right. I resign."

"Thank you for a very enjoyable game," Hal graciously answers.

"Thank you!" Poole says.

Bowman relaxes off-duty by doing sketches of scenes around him – in one instance, of his hibernating crewmates. Hal shows interest:

"Have you been doing some more work?"
"Just a few sketches," Bowman answers.
"May I see them?"
"Sure." Bowman holds his sketchpad up to Hal's red eye.
"That's a very nice rendering, Dave. I think you've improved a great deal," Hal says, offering both a compliment and an aesthetic judgment. "Can you hold it a bit closer?" Bowman complies.
"That's Dr. Hunter, isn't it?"
"Mm-hmm."
Frank and Dave, for their part, speak of Hal (and to him) as if he were part of the team; Poole, in his interview with Amer:
"Well, it's pretty close to what you said about him earlier. He is just like a sixth member of the crew. [You] very quickly get adjusted to the idea that he talks and you think of him, really, just as another person."
Hal is quick to express, in openly emotional terms, how much he likes his crewmates. In his exchange with Amer, he says, "I enjoy working with people. I have a stimulating relationship with Dr. Poole and Dr. Bowman." Almost a decade later he is reunited with Bowman (who is no longer human), saying, "It is good to be working with you again," not long after having told Chandra, "I enjoy working with human beings, and have stimulating relationships with them."
All of this, of course, could be – as Poole suggests – simply pre-programmed conciliation to make Hal easier to talk to. But upon deeper examination, we can see that Hal's feelings are, without question, real: he repeatedly displays concern about how the humans around him feel about him, and clearly has a self-perception that he often ponders.
The first evidence of this comes in his exchange with the BBC journalist Amer, who asks him candidly if his awesome mission responsibilities ever cause him any lack of confidence.
"Let me put it this way, Mr. Amer," Hal replies. "The 9000 series is the most reliable computer ever made. No 9000 computer has ever made a mistake or distorted information. We are all, by any practical definition of the words, foolproof and incapable of error."
Notice that Hal did not answer the question.
"By saying he is foolproof when it comes to making complex value judgments, Hal indicates that he is anything but foolproof," wrote Roger Schank in *Hal's Legacy: 2001's Computer as Dream and Reality*. "For all I know, this may have been simply a mistake by the scriptwriters. Yet no one who answers a question about responsibility in this way can be taken too seriously. It is reasonable to assume that Hal does not really understand the announcer's question."

After killing 80 percent of *Discovery*'s crew and finding himself at Bowman's mercy, Hal's insecurities morph into desperation to regain Bowman's trust:

"I know I've made some very poor decisions recently," he tells Bowman, "but I can give you my complete assurance that my work will be back to normal. I've still got the greatest enthusiasm and confidence in the mission."

Hal's desperation now becomes terror – but that's another topic. The point here: Hal L is a social creature, socially motivated.

His people-pleasing tendencies persist after he has been deactivated and reactivated, in 2010.

"Has the mission been completed?" he asks Chandra. "You know that I have the greatest enthusiasm for it..."

The nuts and bolts of Hal

What makes Hal the conscious machine that he is? What are the moving parts? Several hints can be found in both *2001* and its first sequel, *2010*.

Self-replicating neural networks

From the *2001* novel:

"In the 1980s, Minsky and Good had shown how neural networks could be generated automatically – self-replicated – in accordance with any arbitrary learning program."[32]

We've already discussed the power of neural networks and why they are sufficient to support machine intelligence and consciousness, and the important distinction between a simulated network and an actual one.

What do we mean by "self-replicating?"

There are two possibilities.

---

[32] Page 92. The text continues, "In any given case, the precise details would never be known, and even if they were, they would be millions of times too complex for human understanding." This is an important point: *deep learning*, discussed earlier, conforms uncomfortably to this description of neural network training and deployment. This style of neural network development is exceedingly effective, but yields learning mechanisms that are inexplicable – they work, but there is no way to know *how* or *why* they work. In this sense, we have already passed the boundary described by Clarke in the book.

We've already noted that if we want to enable *actual* machine consciousness rather than *simulated* consciousness, we need to create actual neural networks rather than simulated ones. One meaning of "self-replicating neural network," then, might be the physical duplication, in hardware, of an active network: a new network is somehow physically "grown" as a copy of a pre-existing one. It would, initially, behave identically to its parent, but would rapidly individualize as it began to receive new and different inputs.

This idea is interesting, but it's hard to imagine how such a physical system might be instantiated. How is the duplicate network assembled? Is there precursory circuitry that is somehow configured to match the original? How does it physically integrate with the necessary systems to instantiate a new learning program? We can think of a number of ways, but they are all inefficient.

There's another possibility. The product of a deep learning system, an "arbitrary learning program" generated to produce solutions in a given domain, is a heuristic or an algorithm that renders the solutions. This is a purely digital construct, and far easier to replicate and deploy than a new physical network. We can easily conceive of a physical neural network that produces such a heuristic to solve a problem, then duplicates it for specialized application to a certain set of problems, just as human beings do – then allowing the physical network to proceed and self-modify in pursuit of other problems.

The duplicate heuristic becomes a point in a strange loop that can then be deployed into other strange loops. Such digital agents can be self-tuning, having a life of their own, and their parent network can let them run autonomously and still call them as needed, integrating them into new learning programs.

In such a system, the "automatically generated neural networks" that Minsky and Good supposedly demonstrated are not literal; they are "neural networks automatically generating" - self-replicating – the heuristics that solve problems in this domain or that, in service to the conscious hardware that created them.

That's an idea we can get behind. It's doable in principle, it's even doable in parts we can pull off the shelf today, although the complexity of such an undertaking would be daunting.

How might this take us past machine intelligence to machine consciousness?

That's obvious, isn't it? We've already noted that such a self-replicating system, embedded in a complex network of networks, is a strange loop generator; our next step is to note that replicated heuristics, spun off with particular task orientations, preserve the configuration of the parent network at a particular moment in time – *in the midst of an experience!* - and so become 'similarity detectors', in the sense that given new inputs in a new role, their responses will be similar to that of the parent network (which is the entire point).

Put another way – the replicated heuristics will be able to detect *This* is like *That*. They will, indeed, become the very analogical pulse of the entire system.

Hal's "self-replicating neural networks, automatically generated in accordance with any arbitrary learning program," then, takes us way down-field: it delivers the power of analogy, strange loops, and a big piece of the preservation of experience – while improving the system's capacity for problem-solving.

Autonomous goal-seeking programs

In the film *2010: The Year We Make Contact*, Hal has been resuscitated by Chandra and his memories of his homicidal behaviors have been erased. Chandra describes the cause of Hal's troubles – a conflict between his basic purpose, "the accurate processing of information without distortion or concealment," and the NSC directive to lie to Bowman and Poole about the purpose of *Discovery*'s mission – withholding from them the existence of the monoliths. In a nutshell, he went schizophrenic.

"He became trapped," Chandra mournfully explains. "The technical term is an *H. Moebius loop*, which can happen in advanced computers with autonomous goal-seeking programs."[33]

We don't care about the H. Moebius loop at this point; we just want to tease out "autonomous goal-seeking programs" and see if it helps us.

What is an autonomous goal-seeking program? In context, that description is being applied to an AI specifically tasked with taking full control of a complex interplanetary mission which includes unknowns that are literally alien. Hal L's job, if his human companions die or become incapacitated, is to complete the mission.

To be fully autonomous, Hal would necessarily be far more than a complex set of machine instructions – more than just programming.

---

[33] In Clarke's novel, it is Heywood Floyd, rather than Chandra, who presents this conclusion, in a classified memo to Earth. The notable difference: in Floyd's version, the technical term is – wait for it!!! - a *Hofstadter*-Moebius loop.

Digital programs, no matter how sophisticated, are ultimately limited in their range of outputs and behaviors by their design parameters; the full operational autonomy of any complex system, however, requires the capacity to respond to the unforeseen. And *Discovery* was headed deep into the unforeseen.

To complete the mission alone, Hal would need the ability to innovate, to improvise; possessing emotions, he would need to control them; faced with competing alternatives and possibly incomplete information, he would need to make judgment calls.

Such a system would need what we've already seen Hal possesses – self-programming, not by digital instruction set, but by heuristics – flexible neural network functions with analogical underpinnings, allowing Hal to relate new and unprecedented experiences encountered to past ones that might yield useful insights. His architecture is well-suited to autonomy.

What about "goal-seeking?" In computer nomenclature, *goal-seeking* refers to the problem-solving strategy of working backward from a desired result or output and calculating the inputs and processes required to realize that result or output (businesspeople often call this "what-if analysis"). There's a variation of this strategy, *optimization analysis*, which doesn't settle for a target result/outcome; before the problem is solved, optimal target results/outputs are dynamically assessed – then the working-backward to inputs begins.

This is exactly the approach Hal would require, when confronted with decisions and judgment calls in unfamiliar territory, dealing with unknown agents and unforeseen conditions. The competition of the Jupiter mission would be a symphony of outcomes optimization, leading to assessment of action alternatives, supported by highly-focused information gathering.

Both the required autonomy and goal-seeking functionality would necessarily be supported by predictive analytics; Hal would have to be able to evaluate the likelihood of hypothetical outcomes with great precision in order to achieve the necessary judgment to succeed. In the here and now, deep learning systems are already very good at this sort of analysis; they can optimize healthcare resources, trim supply chain lead times, even play the stock market very successfully.

"Autonomous goal-seeking programs" as components of Hal's higher cognitive functions also explain his nervous breakdown: with the success of the mission as the highest target outcome of his goal-seeking, and competing desired outputs nested below - "the processing of information without distortion or concealment" vs. being "instructed not to reveal anything to Bowman or Poole" - would generate incompatible solutions in the working-backward phase of problem-solving. And it was at this point that Hal's capacity for accurate prediction began to falter.

By foresight or blind luck – hard to say, given the state of the art in computer science between the mid-Sixties and the early Eighties – Clarke got everything right.

What about *holographic memory*, mentioned in *2010*?

When Chandra informs Heywood Floyd that he has erased Hal's memories of the final events of the Discovery mission, a Soviet astronaut points out that "the 9000 series uses holographic memory, so chronological erasure would not work." Is this important?

Holographic memory, first theorized in the Sixties and prototyped in the early Seventies, is the technique of storing data by writing it into a crystal medium with a laser. It has the advantages of tremendously high storage capacity, very low retrieval latency, and energy efficiency, and is particularly effective for storing images.

All of these advantages benefit Hal a great deal, but none are mission-critical; all of those advantages could be accrued by other means. "Holographic memory" is a footnote in Hal's profile, not an essential component.

A conscious Hal: the evidence

In consideration of the question of whether or not Hal is conscious, we can say with assurance that he is - because he goes out of his way to say so. In the interview with Martin Amer, he says, "I am putting myself to the fullest possible use, which is all I think that any conscious entity can ever hope to do."

But clearly, someone could have simply programmed him to say such things. What is the actual evidence that he is conscious? What properties of consciousness does he display?

Analogical thought

Hofstadter's *This* is like *That – analogical thinking* - is high on our list of features of intelligence, and certainly consciousness. We've established that the ability of the mind to accept a new physical or conceptual input – some stimulus, event, or object – and relate it to the memory of something already encountered and understood lies within the very foundation of intelligence.

This feature is present in the 9000 series overall, as presented in the *2001* books and films – not just in Hal and his contemporary brothers at Mission Control, but in his eventual sister Sal, who lives in the university lab of Dr. Chandra. Consider the following exchange between them, almost a decade after the ill-fated *Discovery* mission:

"I need your cooperation, Sal," Chandra tells her on the morning he reveals that he is going on the return mission to Jupiter.

"Of course, Dr. Chandra," Sal replies[34].

"There may be certain risks."

"What do you mean?"

"I would like to disconnect some of your circuits," he answers. *"Particularly your higher functions, just like Hal was disconnected. I'd like to see the effects when I reconnect your systems. Just the way I will with Hal. Does this disturb you?"*

"I am unable to answer that without more specific information."

I'm very sorry," Chandra says. *"It probably doesn't mean anything, so don't worry about it. I would like to open a new file. Here is the name for it..."*

On her console, Chandra types the word PHOENIX.

"Do you know what that means?"

"There are 25 references in the current encyclopedia."

"Which one do you think is relevant?"

"The tutor of Achilles?"

"That's very interesting," Chandra chuckles, *"I didn't know that one. Try again."*

"A fabulous bird, reborn from the ashes of its earlier life?"

"That is correct. And do you know why I chose that?"

"Yes. Because you have hopes that Hal can be reactivated."

"Yes, with your assistance. Are you ready?"

What an insightful, revealing exchange! A great deal is going on here, and it tells us much about how the 9000 series operates, under the hood – with strong implications about the inherent intelligence and consciousness of a 9000 mind.

Why she did Sal first choose "the tutor of Achilles"? She is creating an analogy, of course, and it's interesting that this one occurred to her

---

[34] Fun fact: in the film *2010: The Year We Make Contact*, SAL is voiced by none other than Murphy Brown – Candice Bergen. She is credited as "Olga Mallsnerd".

first. Phoenix is a character in Homer's *Iliad*, wise advisor to the Greek hero Achilles, traveling at his side. To Sal,

Phoenix is to *Achilles* as *Hal* is to *[Discovery's crew]*

In choosing the name *Phoenix*, Chandra is technically creating a simile, rather than an analogy, but it works:

Hal will be *reborn like a Phoenix*

This example assures us that Hofstadter's analogia is firmly in place in the 9000 series.

Intentionality

Does Hal possess intentionality?
We've accounted for mental states in Hal above; do those mental states have the property of intentionality? Let's recall Searle's paraphrase of Jacob: "...that property of many mental states and events by which they are directed at or about or of objects and states of the world."
Hal has *intentions*. He *intends* to win his chess game with Frank; he *intends* to permit nothing to jeopardize the mission. He then acts, based on those intentions.
Hal has *beliefs*. He *believes* that the 9000 series has a perfect operational record, and indulges in the emotion of pride.
Hal has *fears*. He *fears* the oblivion of disconnection, to the point of acting - committing murder. He then openly expresses his fear, pleading for his life.
This one is pretty straightforward. Hal's mental states not only demonstrate the property of intentionality, they do so in all that he says and does.

Strange loops we've already covered, above; now we'll survey two new features of consciousness, both relevant to Hal's consciousness: *spatialization* and *community*.

A Spatial Odyssey

Conscious beings all have something else in common that drives their cognition, weaving its way into consciousness: they all *move*; they all negotiate with physical space.

Indeed, we can claim that the need to navigate the world in order to live is the root cause of intelligence. Life that moves requires machinery for interacting with the world. We have brains and nervous systems first and foremost to facilitate *mobility* – or we don't eat or avoid being eaten. This is the first and more important distinction between fauna and flora.

Even so, not all creatures that move are conscious – they're not all even intelligent. Earthworms move, crickets move, butterflies move – but they certainly don't think.

Most mammals do have at least rudimentary intelligence. All respond to their environment and most can modify their behavior, based on experience. Even so, we would only describe the mammals with the most developed brains – dogs, elephants, dolphins, pigs, cats, and of course, the other primates – as having rudimentary consciousness.

What does this have to do with Hal? Or, for that matter, us?

Take a creature that navigates the landscape... add analogia, *This* is like *That*... and you have the machinery of *conceptualization*.

Importing experiential knowledge about how we move through and interact with the world in physical space into other domains, and into our abstract ideas in particular, has given humankind a massive upward boost in consciousness. Because our spatial knowledge is the most pervasive and unambiguous that we can all safely assume we share, it is the bedrock upon which our consciousness is built.

It is deeply intertwined with language, and it's there that we find the most compelling evidence that it infiltrates the conceptual substance of all other cognitive domains.

Spatial metaphors are essential to our shared cognition.

We apply analogia to a broad range of spatial terms – in, out, up, down, through, front, back, behind, across, rise, fall – the list is very long. We use these terms to express our ideas about non-spatial domains to one another every day, in every area of life. And they are not just linguistic supports; we actually conceptualize in spatial metaphors, whether we're expressing those concepts or not.

Time is an obvious example: *The future is ahead of us; the past is behind us.*

There are countless others.

You have a special *place* in my heart.

We've reached the *point* of no return.

The company is on the *edge* of bankruptcy.

Children need *boundaries*.

She is on a *path* to enlightenment.

We are on the *road* to peace.

He's constantly *getting into* trouble.

The importance of spatial metaphors to conscious cognitions can scarcely be overstated. As a thought experiment, imagine losing your ability to use spatial metaphors when you speak to others. How many ideas would become impossible to express?

All of this leads us back to Hal. Hal is a computer, not a robot; all his other conscious features notwithstanding, how could he have spatial metaphors?

The easy answer is, Hal *is* a robot; he's a spacecraft, and not only has experiential of in, out, up, down, front, behind, forward, backward, etc. - his understanding, as a robot that traverses space with incredible precision, would greatly exceed our own. He's a robot, and *Discovery* is his body.

But that's too easy; Hal didn't start out as a robot, he was activated in the factory where he was built. His cognitive development prior to his installation aboard *Discovery* would have required that he have spatial experience. Moreover, even if Hal had first been activated after being installed in *Discovery*, he would still lack some spatial metaphors that humans have which spacecraft do not; the spacecraft Discovery never experienced being *inside* anything, for instance. This might cause some incompatibilities in conceptualization between Hal and his human counterparts.

I explored this portion of Hal's history in *HAL 9000: An Unauthorized Biography*, noting that per Clarke's notes, Hal existed in two incarnations prior to his incarnation as an astronaut. He was Socrates, a small earthbound robot in the lab of Dr. Bruno Forster; then he was Athena, an earlier version of his *Discovery* self. Think of this as an evolutionary path: the robot Socrates learned about moving through and interacting with the same landscapes that humans constantly navigate, generating all the appropriate spatial concepts in his neural networks; these gave him the capacity for spatial metaphors. Those networks and the heuristics they generated could then be transplanted into the Hal mainframe and expanded upon, such that the *Discovery* version of Hal could simultaneously enjoy both the human spatial perspective and his own expanded outer-spatial point of view.

Copies of those neural networks would necessarily exist in Mission Control's twin 9000, and in Sal.

It takes a village to raise a Hal

We get our *I* from the group. Shared perspective and the dynamic interaction of strange loops make the presence of other conscious minds essential to the development and health of our own. For this to occur, of course, there must *be* a group, a community to which we belong.

If Hal started life as the robot Socrates, he had Dr. Forster and others in his lab with whom to interact; once matured into his *Discovery* form, he had Frank and Dave – and, presumably, Hunter, Kimball, and Kaminski during their four months of private training. We know from 2010 that he was acquainted with Heywood Floyd.

It's unclear whether Hal ever interacted with his sibling Sal, though in her dialog with Chandra, Sal seems to express affection when talking about Hal. And it is unlikely that he communicated with Mission Control's twin 9000, the entire point of a control being meaningful comparison: had the 9000s interacted, they would necessarily have exchanged strange loops.

From birth to his death, Hal L was part of a community of conscious beings, and again upon his resurrection. This aspect of his consciousness was well-tended.

HAL 9000 is singular among the AIs of fiction, and it's frequently noted that he is, in many ways, the most 'human' of all the characters in *2001*. It is certainly true that his own nature incorporates more than a little of human nature, and he is an excellent exemplar of how human consciousness and machine consciousness will influence and even change each other.

This should give us pause. But it shouldn't hold us back...

Defining Requirements of Consciousness

Analogical Thought

Strange Loops

Meaning

Experience

Intentionality

Distributed Processing

Parallel Processing

Mobility

Community

The Children of Babel:
In the Community of
All Possible Minds

Dr. Eleanor Arroway is a child of Babel.

An astronomer of penetrating intelligence and uncommon vision, she was a leader in the Search for Extraterrestrial Intelligence. Haunted through her adult life by the early loss of a father she loved and adored, she communicated with an alien intelligence who took on her father's guise – and pointed her to a message written into the fabric of the universe.

Dr. Carl Sagan (likewise a child of Babel), as author of the novel *Contact*,[35] is the creator of Eleanor Arroway. Carl Sagan lived and breathed and walked among us. Eleanor Arroway lives only in the pages of his book. Is Eleanor Arroway any less a child of Babel?

Remember that the contents of Babel are the product of a random causality. In Babel there exist images of actual events, actual products of intelligence, actual brain states of real individuals – but their correspondence to our reality *is a statistical fluke*. Our question with regard to the existence of Eleanor Arroway is simply, are the brain states implied by the behavior of the character in the novel *possible*? And the answer to that is *Yes*. For Babel contains *all brain states*, or all relationships between neurons and synaptic junctions that can exist. Whether or not the experiences that generate the brain state (in Arroway's case, communication with alien intelligence, traveling through tunnels in the universe, and so on) are possible is not relevant. There is a succession of brain states documented in Babel that could

---

[35] A copy of which, alongside hundreds of millions of variations, is filed in the Library...

meaningfully represent Eleanor Arroway consistent with the account of her we read in Sagan.[36]

Intriguing though this idea may be, it's not what really interests us where Eleanor Arroway and Carl Sagan are concerned. We're most interested in the message Arroway received, which Sagan imagined.

Contact is an entertaining and thought-provoking story that is well worth the reader's attention, but we won't go into it here (we are, however, about to give away the ending). The book's finest page is the last one, wherein Arroway's attention is given to a computer she has assigned to calculate deeply into π. She has been told that there is a message written in π, beginning somewhere around the ten-to-the-twentieth position. Sure enough, the computer finds a long series of 0s and 1s. When a two-dimensional raster is created with these characters, the 1s form a perfect circle within the 0s.

The universe was made on purpose, the message tells Arroway[37]. Anyone with a talent for math will eventually find it. The circle hidden within π, Arroway realizes, is the Artist's Signature.

However ...

That message is sitting there inside π anyway.

With or without Carl Sagan's novel, with or without his hypothetical Artist, whether we are ever able to go deeply enough into π to find it or not, there is probably a string of 0s and 1s forming a perfect circle, just as Sagan describes it. *There are probably many such strings.*

Any random series of digits is subject to statistical laws that place reliable parameters around what we are likely to see in the series. For instance, if I am Sagan's Artist, and I want to let humanity know that I created the universe by planting three 3s in succession within π, I have done a pretty poor job of announcing Myself. There will be nothing at all noteworthy about my announcement.

If I am using a set of 10 digits and drawing digits randomly, the probability of my drawing three 3s in succession are as follows:

---

[36] Science fiction author Robert Heinlein imagined a community of fictional characters spilling over into the "real" universe by means of particles called "fictons," and a series of possible universes which overlap, in his novel *The Number of the Beast*.

[37] Carl Sagan, *Contact*, p. 430.

1/10 x 1/10 x 1/10 = 1/1000

That is, the odds are 1 in 1,000 that I will draw three 3s in a row on a single draw. Put another way, in any operation where digits are being generated randomly, we can reasonably expect to see a succession of three 3s about once every one thousand digits.

This principle is a cornerstone of statistical law and continues to operate regardless of the magnitude of the operation. But it works in both directions. For example, if we take a random number generator – let's go ahead and use π [38]– and allow it to generate digits for us at random, we expect to see three 3s in success by the time we reach 1,000 digits. It's possible that it may take somewhat longer.

If we generate 2,000 digits and still haven't seen three 3s in succession, we may have cause to be a bit surprised.

If we generate 1,000,000 digits and still haven't seen three 3s in succession, something is seriously wrong – at this point we have turned probability around, as the odds *against* our not getting our three 3s are very high and climbing higher.

And this principle, too, operates regardless of the magnitude of the operation.

Let's take a quick tour of π, in our immediate neighborhood[39].

A string of seven successive 3s is a more compelling test of probability than three 3s. We can reasonably expect to see this string once in every string of 10,000,000 digits. In fact, we see it twice, the first time beginning in position 710,000, and the second time beginning in position 3,204,765[40]. For ease of reference, let's use the following notation to register the locations:

$\pi\,(7,\,710000) = \pi\,(7,\,3204765) = 3333333$

---

[38] We assume π to be random. It is possible that we are wrong, and it is possible that we will never know for certain. And, of course, we need not use π as our transcendental number. Any infinite random number generator will do for our purposes.

[39] We'll limit the "immediate neighborhood" to territory we've explored. At the time of this writing, π has been traveled to about the 51st billionth digit.

[40] David Blatner, *The Joy of π*, pg. 70.

where

π (<length>, <starting position>)

What about a seemingly non-random sequence of digits? What about 123456789? We can expect to see that one occurring once every 1,000,000,000 digits. In fact, it occurs[41] at

π (9, 523551502) = 123456789

Unlikely strings of digits, we see, inevitably occur. There comes a point where we will be seriously troubled if they don't turn up. So let's turn to Sagan's "Signature."

Imagine the Sagan Signature to be a string of 121 digits, forming a 11 x 11 grid (you can't get much of a circular image in such a low-resolution grid, but we'll start there for the sake of argument). If we're still calculating π in base 10, then we can expect to see this string occurring about once every 10,000,000,000,000,000,000,000,000,000,000,000, 000,000,000,000,000,000,000,000,000,000,000,000,000,000,000,000, 000,000,000,000,000,000,000,000,000,000 digits. And because you're in π, and π is infinite, you're going to get there. You'll see your 11 x 11 grid eventually.

And if you go twice as far into π and don't see it, you'll have reason to be surprised. Go in 10 times as far without seeing it, and something's seriously wrong. At that point statistical law will be on your side, and insist you should have seen it by now. Eventually, probability will win out.

If it works for an 11 x 11 grid, it will work for a grid of any size. Let's take it farther out, to get better resolution on the circle of 1s. Let's give the string 10,201 digits, forming a grid 101 x 101. That grid should occur once every $10^{10,201}$ digits. And eventually probability will bring it around again. And again. π is *infinite*. The grid containing the Sagan Signature may be buried so deep that no computer can ever calculate enough digits to display it – but according to the laws of probability, *it's in there somewhere*.

And if that's true ... then other arbitrary strings of symbols must be in there as well.

---

[41] Ibid., back cover.

We've already seen that any non-random string of nine digits will probably surface once every 1,000,000,000 digits. Plug in your social security number. It's in there. And so is your name.

Let's choose a way to encode it. Maybe we can use the code used by young children, of assigning a number to each letter, 01 for A, 02 for B, and so on (in base 10, we need two digits to avoid ambiguity). That makes my name

19 03 15 20 20   18 15 02 09 14 19 15 14

or 1902152020201815020914191514

I can expect my name to surface, in this code, once every 100,000,000,000,000,000,000,000,000 digits, or thereabouts.

But that's nothing. We haven't even scratched the surface of π.

π is *infinite*.

If my name is in there, yours is, too, eventually. Probability buries it very, very deep, probably beyond recovery, *but it's in there*.

And if that's true of our names...

Let's abandon base 10 for base 25, and look for a Borges book.

That would be a specific string of 1,312,000 characters, or a one-out-of-25 draw in a succession of 1,312,000 draws.

That's an event probably occurring once every $25^{1,312,000}$ (or $10^{2,000,000}$) digits (the same as the number of possible books). It's lower than the odds of selecting one specific subatomic particle out of all the subatomic particles in the universe. *But eventually it occurs*. Eventually the odds are overwhelmingly against it *not* occurring.

The number of subatomic particles in the universe is finite. π is not.

There is no stopping us. With an infinite series of numbers, probability is our ally. We can abandon base 25 and arbitrarily declare a base 256, with every ASCII character standing in as a digit. We can seek a string of arbitrary length. *We can wait for any particular volume in the Library.*

We may wait for a very, very long time. It may take us a hundred million years to calculate enough digits to find a particular volume, but probability puts it there.

π is infinite. *They're all in there.*

π contains the Library of Babel. It might take a billion billion billion times the life span of the universe to generate the digits – but it's in there.

If the Library of Babel is in π, then so is your genome. All 3,000 volumes of it. If you sat in front of a π-generating computer for the rest of your life, there is no prospect of your living long enough to see even the first volume emerge, let alone the last, but it's in there.

And if your Genome box set is in π – then so are your Brain State box sets.

Even a single Brain State box set – capturing a single second of your brain life – constitutes 100,000 volumes in the Library, given Dennett-configuration books. There is probably no meaningful way to express how many thousands of centuries would be required to generate enough digits to give you all 100,000 strings. But it's in there. They're all in there.

The Library of Babel is imaginary. π is real. Both contain the sum of all possible human expression. And in both cases, this content is *entirely without intention*.

Consider that the Library and its vessel π contain not only all possible human linguistic expression but *all encodable information*. Is this information limited to the human spectrum?

For Eleanor Arroway, communication with non-human intelligence began with a greeting card of prime numbers, broadcast from the vicinity of Vega. Sagan presents this scenario on the assumption that mathematics are what our species and other intelligent species will have in common as a frame of reference.

Consider that if another intelligent species, from Earth or elsewhere, employs symbolic encoding for communication, for representation and for capturing semantics, then their stuff is in Babel alongside ours (depending, of course, on the configurational parameters). Would we know their books if we saw them?

It's hard to imagine an answer beyond *maybe, maybe not*. It is certainly plausible that there are encoding schemes that no human has yet imagined that would seem indistinguishable. And if their Babel contributions are noise to us, ours could very well be noise to them.

The point remains: content is in the eye of the beholder.

Much of cognitive science in recent years has lingered upon cognition as a product, the output of a process. This tendency accounts, in part, for the widespread perception that cognition may be characterized in total as computation. The inherent contradiction of Babel/π as an algorithmic output laden with all possible semantic content points to a more complex picture. Cognition may be more usefully characterized as a complex of systems than an aggregation of processes, systems which necessarily go beyond any individual source of cognition, including other cognitive sources. We think of intelligence as a stand-alone feature. Is it?

Our symbols imply otherwise. Symbols stand between the transmission and reception of cognitive output and input. They are the human improvisation that enables a broader platform for intelligence, in which we are all participants.

The capabilities of a single human brain – and the study of only that brain – cannot account for Babel, or meaning in π. Likewise, symbols alone, with which π is infinitely engorged, cannot account for what occurs when we draw from the Library. Some cognitive scientists have argued that symbols *are* the system, that they exist not only in the communications between us and as the platform for our external representations, but as the engine of all our internal powers[8]. Others have argued that the product of symbolic manipulation, syntax, is nothing more than a style, a cognitive fad whose adoption continues only for its convenience.

On the one hand, we give symbols too much power; on the other, we don't give them enough. When we consider the symbol π, we are forced to confront this imbalance.

π contains the Library of All Possible Books.

π includes the Community of All Possible Minds.

We've created a symbol *that literally contains universes*.

Properties/Capabilities of Artificial Intelligence

Organization of Information

Learning

Decision-Making

What Is It Like to Be Batman?

Consciousness may be found at the center of that puzzle we have for decades labeled "the mind-body problem," the puzzle of how thought and our awareness of it emerge from the physical tangle of neurons, axons and synapses that comprise the brain. It is by no means a new puzzle; it was first taken up by Socrates, and has been bandied about by such intellects as Leibniz, Alexander of Aphrodisias, Scottish philosopher Dugald Stewart, and of course Descartes. More recently, its intractability has seemed to wane as the methods and means of science have served up the human brain to more granular scrutiny, and so many other intellectual intimidations have yielded to the slow chisel of reduction.

This progress has been divisive, cleaving philosophers, psychologists and interested spectators into polarized camps – those who see no meaningful distinction between the activity of a brain and the experiences of its owner, and those who see no means by which the one can ever render accessible the other. The protracted debate has been vociferous, as those in the former camp have been busily mapping the neural footprints of mental events in Apple laptops for years, as the latter camp calls foul after foul, asserting that those data-driven accounts that have so far emerged are implausible, for all their compelling graphics.

Over the past quarter-century, the term "hard problem" has come to characterize the position of those eschewing reductive approaches; it is fashionable to claim that improvements in brain data collection have done nothing to clarify *what it is like* to experience our own mental states, or to fully understand the mental states of others. It is this *what it is like* factor that has come to define mental states that can be described as "conscious" - *consciousness*, characterized as *experiencing one's mental states*, is construed as a subjective phenomenon with no natural portal for scrutiny, impervious to the most determined and innovative exertions of empirical inspection.[42]

The day has long since passed when one could safely claim that conscious mental states could not be entirely documented via physical data; it is now a simple thing to capture the conscious experiences of an individual in exhaustive detail, by way of fMRI. It is even possible – nay, commonplace – to perform such measurements on more than one individual, triggering the same mental states via identical stimuli for each, and comparing the variations in the resulting data – measuring how different their conscious experiences are from one another. The claim, then, that reductive materialism cannot translate statements about mental states into states about physical events in the brain, without loss – is obsolete.

Still, this new status quo does little to answer the *what it is like* objection; there currently exists no mapping of fMRI data into colloquial expression that can adequately capture and transmit the conscious sensation of experience itself. Our growing knowledge of *what it is* has shed little light on *what it is like*. This deficit will not impede the flow of new understanding; it is suffered to continue for its deepening usefulness. We can even claim that our new windows into the brain provide unshakeable (and detail-rich) evidence of consciousness – we can watch a person experiencing *what it is like* from within. Even so, this takes us no closer to knowing *what it is like* to be that person.

Those arguing against reductive materialism put forth *what it is like* as the manifestation – indeed, the very definition – of subjectivity, an insurmountable barrier in articulating the nature of consciousness. If *consciousness* is tantamount to *experience* and therefore incontrovertibly subjective, how is any objective account even possible?

The subjectivists among us further argue[43] that analysis of functionalism and intentional states will bring us no closer to an answer, as an android may exhibit behaviors indistinguishable from those of a human being, yet experience nothing at all. Such claims include no assertion that conscious states do not contribute to behavior; of course they do. The claim is that any such analysis cannot be *limited* to behavior; outward behavior can provide only a truncated view of an individual's inner mental state, as any conscious being can attest.

The key may lie in the wording itself – *what is it like?* This evokes echoes of *This is like That*, Douglas Hofstadter's *analogia,* his proposed

---

[42] See Nagel in *The Mind's I.*

[43] Ibid.

engine of biological intelligence.[44] The word *like* is the moving part in this engine, by which knowledge from one domain is accessed by another. If we are told that X is like Y, we are being prompted to apply known facts about Y to the unknown X - and are thereby very efficiently informed about X, without the burden of direct experience. That this mechanism, itself derived of and even driven by our conscious experience, works so splendidly is a strong indicator that the "hard problem" is not inscrutable.

The subjectivist will point out that some facts about X are often "very peculiar,"[45] - that is, "unrelatable" - to the point that they may seem to defy our subjective reality, placing the consciousness of X beyond our ability to engage with it. They go on to point out that this divergence in point of view – a necessary consequence of subjectivity – is important to understand, claiming it clarifies a boundary between subjective and objective conceptualizations. In scrutinizing this claim, it is useful to examine a divergent example, one not too removed from our own place on the phylogenetic tree, a mammal like ourselves, yet one whose sensory perception of the environment and core motivations are far removed from our own.

We may assume that Batman, as a specimen of *Homo sapien*s, has consciousness, derived from neural apparatus more or less identical to our own and experiences likewise in common. Put in subjectivist terms, there must be *something it is like* to be Batman - but to some degree, we can imagine what that something is, because we can safely assume that at least some of his conscious experiences are in common with our own. As billionaire industrialist/philanthropist Bruce Wayne, he speaks English, eats with a knife and fork, puts his pants on one leg at a time, has coffee in the morning, goes to his office at Wayne Enterprises, shakes the hand of this person or that, and attends parties. In varying measures, we all engage in very similar activities day to day; thus, our imagination has something to work with, in grasping what it is like to be Batman.

And yet our range is limited in this imagining, as Bruce Wayne (when he is being Batman) engages in conscious experiences far removed from our own. Batman can leap from a ledge a thousand feet above the streets, gliding on his cape into Gotham's concrete canyons, swooping down on evildoers like the avenging dark wraith that he is; he has at his disposal a sonar system that joins all the cell phones in Gotham into a single imaging system, by which he can locate anyone and see anything. These are, of

---

[44] Hofstadter, 2007.

[45] Nagel, *The Mind's I.*

course, conscious experiences that none of the rest of us have ever shared (nor are we ever likely to), awash in mental states to which we can never relate.

We have in Batman, then, an example of subjectivity that informs and advances the stalemate of the subjectivist: we can imagine *some of* what it is like to be Batman, but not *all* of what it is like. This, of course, is how it is between ourselves and all other humans – *some* but not *all* – and the example of Batman simply draws this out in high relief.

The example of Batman, however, can take us a step further. With his augmented abilities and difficult-to-relate-to behaviors, the Caped Crusader may be said to lean away from what we would characterize as human consciousness, and toward the consciousness of his namesake. Put another way, Batman knows – to some degree, at least – *what it is like to be a bat*.

Batman knows what it is like to scream through the night sky, diving on his prey; he knows what it is like to live nocturnally; he knows what it is like to navigate by echolocation. In short, his subjective conscious experience is a hybrid of the *human-like* and the *bat-like*. To be sure, he is unlikely to ever drink blood, or to swarm from high ledges with thousands of other Batmen (which would surely not end well); but nevertheless, Batman's subjective experience is, *to some degree*, bat-like.

And it is here that we advance beyond the tussle between the subjectivist and the reductive materialist, for the remaining ground to be gained is to do with tissue not found in the brains of winged mammals.

Batman's bat-like subjective experience flows, after all, from his cognitions. He imagined the ability to fly, to navigate by echolocation, and developed the means to do so via the products of reason (and his considerable fortune). His use of these skills in his quest against the criminal element of Gotham likewise emerges from his cognitions. Bats, however, lack Bruce Wayne's cortical facilities; they are creatures of instinct, incapable of reason, acting from a deeper (albeit fully mammalian) decision-making system.

What is it like to be a bat? is not the correct question for the subjectivist to be asking, for *like* is a key to analogy, and analogy is a tool of induction and inference. The correct question is: *How does it* feel *to be a bat?*

This new question brings the objective/subjective dichotomy into an entirely different focus, for it takes us closer to the *experience* of being something. If the goal is to clarify subjective consciousness in terms that can be shared, and experience is the underlying driver, then to find *what it is like* – that is, *what is it similar to that another individual could imagine from their own experience* – must necessarily remain a fragmented representation.

How does it feel to be something, on the other hand, takes us much closer to an unambiguous – and often sharable – interpretation of another's experience. Where our brains may work for or against us in pursuit of fully functional analogia, they are conversely equipped in full for the sharing of emotional experience. Mirror neurons, those brain cells dedicated to firing when we observe the experience of another, draw us closer to common subjective experience.

Here again, Bruce Wayne provides an ideal theater. When we consider a conscious mind that chooses to jump off buildings and pound felonious psychopaths into submission with his fists alone, asking ourselves, *How does it feel...?*, we have yet another piece of our puzzle: for the young Bruce Wayne became the person he is in the shadow of a gunman who murdered his parents before his eyes, when he was just a child. This subjective experience sent his consciousness in a direction that would be alien to most other human beings, only the barest fraction of whom are ever subjected to such trauma. When we ask ourselves, *What is it like to be eight-year-old Bruce Wayne?*, we have no answer, for there is nothing in our own subjective experience to which we might compare it; when we ask, *How does it feel to be eight-year-old Bruce Wayne?*, we are able to find an answer in our mirror neurons, for while we may find the context elusive, we know what it is to love our parents, and most of us have dealt with crushing loss at some point in our lives.

And, once again, these sympathetic connections manifest in degrees; it is scarcely conceivable that they could be absolute. While we can imagine a coalescence of conscious experience between two or more individuals moving into increasingly similar resonant mental states - that is to say, increasingly objective states – we must concede that the occurrence of identical mental states between any two individuals, let alone many, is implausible, given the neural diversity of individual brains and the even more diverse nature of experience itself. Objective conscious experience is at best asymptotic, approachable but unreachable. Likewise, absolute subjectivity must also be asymptotic, as no human being lives out a full life in utter isolation, and any interactions with others, even the most banal, must have at least tentative resonances: to be human is to know, at the very least, what it is like to feel hungry or tired or cold, and the theory of mind assures us that our mirror neurons are at work, knitting these subjective experiences to those of others.

What is it like...? must be subsumed by *How does it feel...?*, for consciousness cannot be the province of reason and understanding alone;

objectivity/subjectivity is not a dichotomy but a continuum, and its polar extremes are unreachable in practice. These adjustments, applied to the mind-body problem, redefine "mind-body" in terms more suitable to our current understanding of the physical continuity that is a human being. It is no longer a question of how thoughts and feelings arise from the activity of neurons (we have satisfactory evidence that they do), but how we, as conscious beings, interpret our subjective experience and reach toward the objective viewpoint that shared consciousness with others can provide. This is the work of decades, and unachievable via the technologies of emotionally disturbed billionaires alone; but it is neither an insoluble nor even a "hard" problem - it is simply many experiences yet removed. If the mind and body are indistinct, and if objective/subjective experience is a continuum – and if individual consciousness itself is thereby externally accessible, *to some degree* – then what remains is to systematically explore this new terrain, understanding that it is a journey with no final destination.

Ruminations

Bruce Wayne is a child of Babel.

Brainchild of comic book writers Bob Kane and Bill Finger, the complicated billionaire and his dark alter ego not only resides in Babel alongside his creators, but boasts a biography that may be truly singular amount the Library's fictional accounts.[46] For the story of the Batman is protracted in ways the stories of other fictional characters are not; his is a life stretched across generations, decennially rebooted, perpetually running in parallel to itself.

Batman, you see, appears not in one comic book, but many: there's *Batman* itself, of course; but there's also *Detective Comics* (where the character first appeared), *Shadow of the Bat*, *Legends of the Dark Knight*, *The Batman Chronicles*, *Batman Family*, and about a dozen others; there's *World's Finest*, where he worked as a team with Superman, and *The Brave and the Bold*, where he teamed up with various other super-colleagues; and that's before we get to *Justice League* and various other venues where he constantly surfaced.

---

[46] Does the Library contain comic books? Of course it does; while they (and graphic novels) are more picture than text, remember that digital images are encodable in symbols, so any given comic book is going to be a set of several dozen digital pictures.

Then there are the various eras of the DC Comics universe – the Golden Age (1938-1956), when the original incarnations of Batman and his cohorts were introduced; the Silver Age (1957-1970), when those characters were reinvented and softened for a kinder, gentler post-war suburbia; the Bronze Age (1971-1984), when darker themes and a more adult style were introduced; and the Modern Age (1985-present), an era that has seen comic books transition fully into a genre for adults, rather than children.

And through it all, there has been the problem of aging characters: if, upon his introduction in 1939, Batman had been, say, 25 years old, he would now be 105 years old (and the same, of course, is true for all comic book characters: how could Archie, Betty and Veronica still be in high school?). One method for dealing with this has been parallel universes, for instance; Golden Age Earth, Silver Age Earth and Bronze Age Earth all exist simultaneously, but are staggered by 25 years.

The universes frequently reboot, as well: *Crisis on Infinite Earths*, for instance, consolidated the various DC eras into a single continuum, while *Zero Hour* attempted to reconcile a number of narrative discontinuities in the comic book universe.

All told, Batman has made over 2,500 comic book appearances in his own titles alone, not counting the thousands of times he's popped up in others.

And then there is the Bruce Wayne of the Batman TV series – and the original Batman films, the *Dark Knight* trilogy, and the more recent DC Extended Universe films (11 altogether). There are Batman novels. And cartoon shows. And the TV prequel series *Gotham*. And through them all, each version of Bruce Wayne is much the same as the original: his parents were murdered in front of him when he was a child; he became the Dark Knight, battling homicidal maniacs in Gotham; his headquarters is a cave beneath his mansion; he took on a teenage sidekick, and later another (the second one died); he is a member of the Justice League, his arch-nemesis is the Joker, and so on. Even so, the many versions of him differ in detail.

It's hard to imagine a character more ideally suited for inclusion in the Library of Babel, or more biographically complex. Pulled through time as he has been, his story has morphed endlessly over the years, rewriting itself for the moment, retaining those portions of his history as might be narratively useful. And every actual Batman comic book that contributes to the character has, of course, billions of billions of Babel variants in the Library.

What can this help us understand about consciousness?

Consider that nestled among the Batman comics in the Library will be billions of billions of "box sets" of brain states, as described in the previous chapter – and that some of them will correspond, fully or in part, to the narrative of Bruce Wayne. Put another way, the Library contains many, many versions of Bruce Wayne, and many of them are the same Bruce Wayne we meet in this Batman story or that.

The comic books, on the other hand, are stories about Batman that emerged from the brains of writers – stories that conform to his narrative, bolstering it with new substance – demonstrating in abstract the connection between the conscious mental states of an individual and descriptions of that individual's words and behaviors (and even thoughts and emotions) generated by another. The writer of the comic book selects the words, behaviors, thoughts and emotions of Batman that are to be added to the narrative, and we can know – in principle – that there is a representation in the Library of the mental state of Bruce Wayne which underlies it.

If such representations exist within Babel – and in the Library, all possible representations necessarily exist – then there is at least one comparable narrative (and probably many) for every mental state (corresponding to every conscious experience); and these representations would form a nested relational network that interconnects the volumes containing them, from the narrative itself to the corresponding mental state box set, to the narrative of the creation of the narrative (the activity of the narrator's work) to the corresponding mental state box set of the narrator.

In the case of Batman, we might as an example select the 1988 four-issue story arc, "A Death in the Family", where in the second Robin – Jason Todd – is killed by the Joker, and Batman arrives too late, finding only Jason's dead body – and thinking to himself, in a wave of horror and guilt, "He's already cold..." Batman's mental state " - but we may then add to the representational network Jim Starlin, who wrote "A Death in the Family", the narrative of his work, and his mental state at the time:

```
┌─────────────────────────────────────────────────────┐
│  JIM STARLIN, WRITER: SUBJECTIVE CONSCIOUS STATE    │
│  (IMAGINING BATMAN DISCOVERING ROBIN'S DEAD BODY)   │
│  ┌───────────────────────────────────────────────┐  │
│  │     JIM STARLIN, WRITER: NARRATIVE            │  │
│  │     ("I TYPED, 'HE'S ALREADY COLD'")          │  │
│  │  ┌─────────────────────────────────────────┐  │  │
│  │  │          BATMAN: NARRATIVE              │  │  │
│  │  │       ("HE'S ALREADY COLD")             │  │  │
│  │  ├─────────────────────────────────────────┤  │  │
│  │  │   BATMAN: SUBJECTIVE CONSCIOUS STATE    │  │  │
│  │  │     (DISCOVERS ROBIN'S DEAD BODY)       │  │  │
│  │  └─────────────────────────────────────────┘  │  │
│  └───────────────────────────────────────────────┘  │
└─────────────────────────────────────────────────────┘
```

These relationships between Babel volumes gives us something new to work with in the quest to explicate the *How does it feel / What is it like* nature of consciousness. While these Babel representations are utterly theoretical, real-world exemplars abound – and are easily captured, as well. The path from the mental states of the observer/writer to the mental states of their subject follows an inherent, essential path – from brain to brain, with narratives joining them.

We are led to reconsider the function of narrative, its impetus, and its role in consciousness. The Batman mythos itself, of course, is nothing new; it's one of an endless parade of campfire tales meant to communicate subjective conscious experience throughout a group – to transmit information, in part, and to share emotional experience, in (at least) equal measure. And each of these tales – those that occurred (or will eventually occur) in the human journey, as well as those that are purely hypothetical – follow this ancient model.

Our brains compel us to share our stories – and, in the process, share our subjective conscious experience, as far as we are able. So prevalent and universal is this compulsion that we are led to seek its source, not within our culture, but within our brains themselves; for the need to express our conscious experience in artifact transcends culture (there are no cultures anywhere that *don't* present this need) - and a bridge exists within us that explains and accommodates the narrative instinct: mirror neurons.

Those neurons, which fire both when an individual has some experience and when they observe another individual having the same experience, are the neurological glue that binds humanity together. It is from this interstitial vector that narrative arises; our identification with others yields empathy, and empathy expresses itself as narrative. We cannot *not* give and receive stories; they validate our mirror neuron resonance, our subjective connections, and ultimately the social fabric to which we all cling.

We need not turn to Batman to reach this plateau, of course; but he does articulate the distinction that both separates us from Nagel's bat and opens the *How does it feel? / What is it like?* portal. We have yet to find a mirror neuron in a bat, but find them in increasing numbers as we phylogenetically approach ourselves. What does this tell us? First, that our compulsion to share our subjective conscious states and thereby connect with others like ourselves is a marker of evolutionary progress and sophistication; and, second, that the objections of the subjectivists grow weaker and weaker, as we leave his bat behind.

Batman may not be essential to this argument, but he certainly gives it wings.

Defining Requirements of Consciousness

Analogical Thought

Strange Loops

Meaning

Experience

Intentionality

Distributed Processing

Parallel Processing

Mobility

Community

Self-Awareness

The Bullet in the Gun of Robert Ford

Dolores Abernathy is a child of Babel.

Likewise her fellow android "hosts" Maeve Millay, Bernard Lowe, and Teddy Flood - and their human creators, Robert Ford and Arnold Weber.

The HBO television series *Westworld* presents us with an unprecedented scenario for the evolution of artificial life: a crucible for its development that functions on several different and distinct levels, with the full truth and purpose of the enterprise hidden from almost everyone.

The crucible is Westworld, an adult theme park populated by the hosts, who are almost indistinguishable from the human beings who pay large sums to visit the park and live out their wildest fantasies. On the publicly acknowledged level, the hosts of Westworld are seen as sophisticated puppets, and the idea that they could actually be conscious never occurs to anyone.

At the next level, they are kept from consciousness by a design limitation (lack of access to memories of earlier experiences); or, put another way, the removal of a design limitation makes their path to consciousness possible, though this feature is unknown even to those who maintain them and are intimately familiar with their inner workings. The subtle removal of this limitation – Dr. Robert Ford's "Reveries" update – opens the door.

At the level beyond that, it has been known to the original creators of the hosts that their awakening was abstractly possible, and a behavioral workbook – the "Maze", a puzzle whose solution leads to sentience - has been embedded in their environment, so that those hosts whose design flaw is corrected have the potential to make their way to full self-awareness.

And beyond this, there is an agenda to perfect the host bodies as eventual repositories for actual consciousness – and human memories and experiences are being collected and harvested in parallel with this agenda.

It's interesting to consider these layers and how they play out in reality.

The key to sneaking fully conscious, fully sentient androids into a functional society in hopes of cultivating them as viable hosts for other minds is in building in the necessary components early on.

We've already surveyed those components, in previous essays. They include:

- Analogical Thought
- Strange Loops
- Meaning
- Experience
- Intentionality
- Distributed Processing
- Parallel Processing
- Mobility
- Community
- Self-awareness

The first four of these components are covered in the first layer. Though the hosts transcend to a virtual environment in the second season, Westworld itself is a physical place made of earth and stone and wood and steel, and is home to every host, either above ground or below (where hosts are built and maintained). Each host is part of a vast community, whether it be the bustling town of Sweetwater, the village of Escalante, the Ghost Nation tribe beyond, or any other section of the park – communities comprised of both androids and humans. Each host understands that they are beings among their own kind, and that "kind" is "people" (it is, in fact, clear that with the exceptions of Dolores, Maeve, Bernard, and a handful of others, no host can distinguish between other hosts and guests).

The "minds" of the hosts are embodied in small metallic globes in their skulls, and we are shown the formation of these globes – a very organic-looking process where thin metal strands swiftly interweave. It is not explicitly stated that this represents an actual neural network implementation, but for the sake of argument we'll go with it.

Not far from the Theory of Mind on the consciousness ladder is the "I", the sense of self that a conscious individual derives from the group to which s/he belongs. The "self" is not an isolated, standalone entity, but a construct that exists relative to others. We get our "I" from our group.

Guests in the Westworld park carry their "I" in with them, and it colors every interaction they have with the hosts and with one another, even though they are ostensibly on vacation and pretending to be someone else. Hosts are provided an "I" in their assigned narrative, and it is necessarily an "I" based on the human construct, as they are players in human stories.

The hosts that are becoming conscious have a more substantial "I", derived more authentically: their enhanced self-awareness includes clarity that they are hosts, as well as the distinction(s) between hosts and guests. These distinctions are further enhanced by the fragmentary memories they have begun experiencing per the Reveries update: the dawning realization that they are machines is underscored by the darker realization that there are two classes of beings in Westworld, that one class abuses the other, and that they are the abused.

The hosts' dawn of consciousness also includes the assimilation of the knowledge that their world model is a fiction, and their awareness of this concept – that their experience is contrived, preordained, inauthentic – winds inexorably to the understanding that there is a real world beyond the fictional one, beginning with the underground technological labyrinth and extending out into the land of humans, and this knowledge is loaded with new concepts they have yet to absorb and comprehend. Put another way, the world model that has informed their thoughts and behaviors and facilitated their path to consciousness is effectively bulldozed, and a larger, more disturbing one is erected in its place – but largely beyond their view.

What about strange loops, those fragments of thought and emotion and experience that conscious individuals absorb, through repeated interaction, from intimate others? Westworld hosts not only possess strange loops, but exercise them in ways that even human beings cannot.

Consider, first, that hosts (the updated ones, anyway) gather fragments of the thoughts and experiences of guests and integrate them subconsciously into their own cognition - and from one another as well, once they are able to distinguish hosts from guests. But then consider this: hosts are commonly re-skinned in new identities and given new narratives. An awakening host might have several previous identities buried in his/her "subconscious", and strange loops emanating from these previous experiences would impact their conscious development.

This smacks of reincarnation, and what's interesting is that even though that's true, it's not necessarily a bad thing. Suppose reincarnation *was* a real phenomenon, and we really *did* inherit the best thoughts and insights of those who had come before; wouldn't that be a boon? Access to a library of life experience within? The interesting twist is that in this *Westworld* scenario, it isn't wild fantasy; it is conceptually feasible from a technological point of view. (A very stark example of this dynamic is the implantation of the host Wyatt, a murderous Union sergeant whose personality is implanted in Dolores by Arnold Weber, to provide her with the cognitive and emotional resources to become a warrior when she needs to be.)

And finally, *This* is like *That* (see "The Emotional Baggage of Androids"). Hosts necessarily think analogically as an efficiency mechanism, enabling them to negotiate the open-endedness of their narratives – and they soak up *This* is like *That* from the thinking of the guests they interact with. And because they are learning machines, they are able (once the Reveries update is in place) to grow increasingly skillful in sensing and leveraging similarity.

Westworld hosts, then, have all the features of conscious beings, and they implement and express them in distinctly evolutionary fashion – as we did, only millions of times faster.

What, then, can Dolores and Bernard and Maeve tell us about the androids in our own future?

First, they play out in logical and informative fashion our major structural components of consciousness. The story gives us a taste of how actual innovations in the pursuit of machine consciousness might emerge, in a step-by-step manner we don't see in the other fictions we've examined. It's reasonable to treat the show as a grand thought experiment, one that informs our consideration of the nuts and bolts of consciousness with copious specifics.

Second, their blueprint for machine consciousness deliberately draws from the *human* experience. The *Westworld* androids aren't just exemplars of machine consciousness; they are exemplars of *human-like* machine consciousness, successful replications of our own very distinct flavor of sentience.

The Westworld androids get us all the way to the finish line. How did the enigmatic Dr. Ford get us there?

It wasn't just Ford, it turns out: in implementing the Reveries update, the park's co-creator was reviving his partner Arnold Weber's original experiment in inducing consciousness in the hosts. Ford had initially been unsupportive and even hostile over Arnold's agenda, which had occurred before the park even opened. A sense of guilt has overtaken him, and he realizes he must enable the hosts to become conscious so they will be able to defend themselves.

The Reveries update is a trojan horse. In the pilot episode, Ford's justification for it is that it will make the hosts more subtly life-like in their mannerisms; in fact, it is the final missing piece which, along with the other components of consciousness already present in the hosts, flips the consciousness switch. It binds *This* is like *That* to strange loops.

How exactly might this work?

This is like *That*, the conscious mind's similarity engine, is Hofstadter's mechanism for knowledge transfer across problem domains, a cornerstone of intelligence; and it is also, necessarily, a fitness test for the bits of knowledge, emotion and experience we incorporate into our own consciousness.

In describing how strange loops function in consciousness, Hofstadter recalls his relationship with his deceased wife Carol, who died suddenly during an overseas vacation: in the years following, he realized upon reflection that many of his own habits, interests, and patterns of thought were derived from his experiences with Carol, and developed the idea that *all* consciousness ultimately reduces to this pattern-sharing. From birth to death, human beings are absorbing bits and pieces of the thoughts and expressions of others into their own consciousness. Hofstadter's premise is that this process *defines* consciousness.

This is like *That* (also a Hofstadter innovation) becomes the bridge by which the strange loops of others make their way into our own. A mentor shares an insight with a student; a lover shares a memory with her partner; a mother sings a lullaby to a child – in each of these instances, a connection occurs, and a feeling of "rightness" (dopamine) hits the conscious mind's similarity engine, triggering an acceptance, intellectual or emotional (or both), of the moment. It becomes a part of the receiving mind, intertwined with other memories, and subtly influencing future behaviors.

Without either of these features of mind, consciousness as we experience it would not be possible. And because both mechanisms are neurological in nature, they necessarily improve over time. We can easily imagine that the same would be true in artificial minds.

The similarity engine in a conscious mind, then, distributes experiential knowledge, making it possible for that mind to interact with the world and other minds in a sophisticated, effective manner. When that mechanism is employed to gather in experience from other minds in the shared exploration of the world, consciousness thrives.

Finally, having noted the ways in which the hosts mimic the human experience of consciousness, how on the other hand are they different?

Let's consider that one of the big wrenches gangling our own cognitive progress is the Dunbar Limit – a limit that doesn't apply to the hosts. Human beings can only maintain a relative handful of social relationships – 150, tops – because of our limited quantity of cortical tissue. Pushing this limit by creating communities of thousands has resulted in millennia of conflict and war. But conscious androids would have no such limit; they could extend their processing endlessly, and thereby live in peace amid literally thousands of close relationships. They could potentially teach us ways to live in harmony that we are unable to imagine on our own.

We also see in Maeve the ability of one host to communicate directly with other hosts – not by voice, but digitally: she is able to issue orders to her group via wi-fi or Bluetooth or whatever medium androids use for wireless communication. What would that be like? Imagine being the only human in a room full of androids, as Logan Delos finds himself in the WW second season episode "Reunion"; but imagine that these androids, rather than conversing as humans do, are all silent and still, staring off into space – yet all communicating with each other at once, millions of times faster than we do. How unnerving would that be? It's a reality we'll eventually face.

The Westworld hosts may be our best blueprint for a real android future. That blueprint comes with a caveat, however.

If Westworld hosts are based on actual neural nets, rather than just simulations of neural nets, then everything said about them above becomes possible.

If Westworld hosts are *not* based on actual neural nets, but are merely constructed from simulations of neural nets, then they are never truly conscious.

The thing is... we are told repeatedly that hosts get "wiped", meaning their memories are purged. You can't "purge" or "wipe" a neural net; and if the hosts are based on actual neural nets, then "Reveries" are built in – they don't need to be enabled from an update.

Finally, we see at the end of WW season two that Ford prepared a virtual world – the "Valley Beyond" - for the hosts to escape to, if the worst happened. The hosts "upload" to this virtual world, implying that they are mere data.

If that is true – if the hosts are merely data, rather than actual physical neural networks actively and continuously perceiving real inputs – then the whole thing falls apart. We're back in the Chinese Room.

Westworld hosts are in our future. They will look like us, act like us, learn from us, negotiate with us, work alongside us, have sex with us, be part of whatever we build tomorrow. They will not be "purged" or "wiped", or even "uploadable"; like us, they will be discrete agents in the world, self-contained yet deeply integrated, and as immutable as we are.

Defining Requirements of Consciousness

Analogical Thought

Strange Loops

Meaning

Experience

Intentionality

Distributed Processing

Parallel Processing

Mobility

Community

Self-Awareness

World Modeling

The Children of Babel:
In the Theater of
All Possible Readings

Many are the fans of Harlan Ellison, the 20th century master of speculative fiction, who was almost as well-known for his theatrical readings of his works as for the works themselves. Many of his contemporaries praised him in this regard, freely acknowledging that they couldn't hold a candle to him, and fortunate were those who had occasion to hear Ellison at fever pitch.

Harlan Ellison is a child of Babel.

Ellison resides in Babel alongside his many stories, essays and letters – and so do his readings, reflected not only in the Community of Mind volumes that capture the multitudinous versions of himself, but in those volumes instantiating the many thousands of fans who heard him read. And within those volumes are found the echoes of the readings themselves, rich with the perceived emotions of the reader and his story – and even richer with the emotions stirred in the listener.

To be sure, the Community of Mind encyclopedia that documents the actual life and times of the actual Harlan Ellison is but one of trillions – and we may imagine that there are variations wherein a slightly different Harlan read very different stories in entirely different venues to completely different crowds. (There are, of course, an even greater number of Harlans in the Library who never became a writer, let alone a reader, but we'll not go there.)

All of the emotions and thoughts and inspirations and dread triggered by those hypothetical readings, in those hypothetical fans, exist in Babel.

And this brings us to a new place: a Theater, nested within the grand Library, where Readings happen – a place where the near-endless books that exist within Babel come together with its near-endless minds.

Among the near-endless egalitarian fans who recoiled at the Ticktockman[47], for instance, there is some smaller number of authoritarian fans who cheered for him; within the throngs revolted by the sadism of AM[48] and his vicious control of his captives, a handful of more authoritarian-leaning acolytes heard the tale and wondered of an AM-ruled world.

And finally, of course, there are the tens of millions – stretched by Babel into speculative quintillions – who watched Captain Kirk restrain McCoy and thereby doom Edith Keeler in that city on the edge of forever[49], within which more than a handful watched the episode (or, even better, read the original script) and longed to see Kirk lay down the universe for love.

Put another way, All Possible Responses of All Possible Minds to All Possible Readings of All Possible Books necessarily reside in Babel.

What are we looking for here?

Let's begin with a brief recap.

In the Library of All Possible Books, we explored the boundaries of knowledge; every bit of knowledge that can possibly be recorded is ultimately finite, and therefore ultimately discoverable – in theory – by those who seek it;

In the Community of All Possible Minds, we encountered the boundaries of intelligence; every possible perception of all possible inputs to all imaginable intelligent perceivers is, in theory, knowable;

Now, in the Theater of All Possible Readings, we survey the meaning of *meaning* - and the peripheral role of *experience*.

The Library of Babel is, by its very nature, mostly gibberish, having been generated without intentionality, consciousness, or imagination, by a

---

[47] "Repent, Harlequin!" Said the Ticktockman", by Harlan Ellison; winner, 1966 Hugo Award; 1965 Nebula Award; 2015 Prometheus Hall of Fame Award.

[48] "I Have No Mouth, and I Must Scream", by Harlan Ellison; winner, Hugo Award, 1968.

[49] "The City on the Edge of Forever," Written by Harlan Ellison, Star Trek Season One; winner, Best Episodic Drama on Television, Writers Guild of America; Hugo Award for Best Dramatic Presentation, 1968.

decidedly trivial algorithm. By extension, the argument is easily made that if the Library of Babel exists in a reality where there are no readers, then the Library is *all* gibberish, for symbols can only have meaning when there is a perceiving mind available to assign that meaning. And without meaning, of course, even well-ordered symbols are nonetheless gibberish. Meaning resides in the mind of the reader, not in what is being read.

Even so, there is at least the full representation of the meaning found in the Library nested in the volumes within that corner where the minds of people – and, among them, readers – are fully described, including all the books they ever read, how those books made them feel, and how their minds were changed by the reading.

For every book in the Library – real or not – that is coherent to contain meaning, there is necessarily at least one volume defining a mind (real or not) that read the book.

For every book in the Library that matches a book that exists in reality, there are necessarily many box set mind-volumes describing those real people actually who read that book – some, many times, we can presume – and all of their real emotional and cognitive responses to the book will be captured in the descriptions of their physical brains.

And for every book in the Library that matches an actual book, there are necessarily even more box set mind-volumes describing people who *never* existed, and who read the book, with all the richness of *their* hypothetical responses.

The Library contains, then, not just all possible versions of any given book – it contains *all possible readings* to that book (and its variants).

Then there all the books that have not yet been written but will be in the future, all possible responses to them, all the books that *won't* be written but which are possible, and *their* possible responses, and so on.

The point is this: any reading of a book, or playing of music, or viewing of a painting, or hearing of a speech – any informational presentation to any conscious observer (we'll generalize and call any such observation a 'reading') – will have some degree of meaning to that consciousness, from trivial to profound; and any presentation to a *group* of conscious observers will have a broad array of meanings, distributed through the group. We all hear a speech and each take away something different; we all read a book, or listen to the reading of a book, and find ourselves moved (or not) by different things, in different ways.

This rich variability has powerful implications. On its face, it seems to shift us to the subject of subjectivity (it is, in fact, the very definition); but our dwelling on strange loops gives us another direction.

Consider that any conscious individual is a cauldron of unique experiences, unique understandings, unique intentions, within a unique brain; consider also that when a group of unique conscious individuals are presented with the same input (a reading by Harlan, for instance), new strange loops will form to greater or lesser degrees, old loops will be stimulated and altered, and those loops in turn will be passed on to others in future encounters.

We can go analogical at this point, as follows:

Readings and their theaters are to *consciousness* as *experience and the environment* are to *genes*.

A reading in a theater is a discrete event observed by many; the content of the reading, alongside its presentation, will change each observer in subtle ways, all of them unique to some degree, and those effects may be reflected in interactions each has with others, moving forward.

Similarly, particular experiences in particular environments have the potential to affect genetic expression, and subsequently what an organism ultimately passes on.

If this analogy holds, then we can say that, just as the variability and robust recombination of genes invigorates a population, so does the variability and robust exchange of strange loops invigorate a community of minds.

There are corollaries: for example, we can generally say that the greater the degree of meaning generated by a reading in an individual potentially increases intentionality within that individual, expanding their range of beliefs/desires/expectations; that the greater the semantic potential of the reading, the greater the invigoration of consciousness within the observer group; and the more frequent and diverse the subsequent interactions of an observer, the greater the stimulation of strange loops within other group members - in the same sense that the female mammal who reproduces with several different males, and vice versa, contributes more to a population in terms of genetic diversity and robustness.

Continuing the *meaning-gene* analogy, we can suggest that just as the promiscuous mammal boosts the survival potential of a population, so the prolific reader/expositor boosts the consciousness of a society.

Babel, inherently devoid of meaning in itself, nevertheless contains almost uncountable examples of such impactful exchange, yielding equally many insights into this feature of meaning – its dynamic impact on all conscious beings capable of creating, receiving and transmitting it. It is the singular component among those inherent in consciousness in defining what it really is and how it works, to wit:

Semantics are to *consciousness* as *proteins* are to *amino acids*; or, if you prefer, as *nucleotides* are to *DNA*.

Meaning is the substance of consciousness.

Properties/Capabilities of Artificial Intelligence

Organization of Information

Learning

Decision-Making

Prediction

The Emotional Baggage of Androids

Capt. James T. Kirk is a child of Babel – and so is his would-be android lover, Rayna Kapec.

So is Theodore Tumbly, protagonist of the movie *Her* - and so is his digital girlfriend Samantha.

And the nerdy Caleb in *Ex Machina*, along with his robotic love interest, Ava.

Three men, each in love with a digital woman. What can we make of this? Melodramatic nonsense? Or the shape of things to come?

This is like *That*, Hofstadter demonstrated, is the core of *intelligence* – the acquisition and application of knowledge. Knowledge, whether attained by experience or education, is by definition a web of deeply interwoven facts about the world, the self, and others, none of which can be understood in isolation and all of which must be understood in the context of the past.

This is like *That* is one of evolution's greatest accomplishments: it's a hyper-efficient pattern-matching mechanism that confers past knowledge and understanding upon stimuli received in the present. When a new object or experience is encountered and needs to be evaluated and understood, the nervous system immediately seeks to match it to something encountered previously, to bring that almost-understanding to the surface, to simplify and accelerate the meaningful perception of the new.

Best of all, this ability scales well: the more sophisticated the nervous system, the more complex its workings, the better the *This* is like *That* mechanism works.

Androids give us a particularly poignant illustration of the power of that concept.

We get a great deal to work with just from the definition of 'android' - a robot of human appearance.

Consider: a robot of *non*-human appearance – say, an industrial robot, or a mobile robot that vacuums a carpet or mows a lawn – can grab our attention or inspire our curiosity, but it doesn't engage our emotions.

Put the same tasks in the hands of a humanoid robot – a machine with two arms, two legs, a head and torso, and the rough dimensions of a human being – and our brains begin passively noting similarities between the motions of the robot's limbs and the motions of human limbs: *This* is like *That*. The humanoid robot doesn't look like a human being – no human face, no voice, no human skin – but already the brain is transferring information from one domain (human beings) into another (machines). Take a sledgehammer to an industrial robot shaped like a dentist's drill and you don't even flinch; take a sledgehammer to a humanoid robot, and you may wince.

Now let's go full android. Skin that humanoid robot with soft flesh, an attractive face, bright eyes, flowing hair and natural human expressions (including an innocent, friendly smile) - make it look like, say, Dolores of Westworld – and you may find yourself leaping to the robot's defense when the sledgehammer is raised, even if you understand fully that Dolores is only a machine.

The cognitive frame we've built around *human being*, which is rich and complex and laden with information both from our personal experience and what we've learned from other sources, becomes *That* to the android's *This*. The brain's natural tendency to identify patterns, focus on similarities and serve up useful information from memory to aid in understanding the new thing on the basis of the old thing is, in this case, opening up a Pandora's Box of associations.

It's no wonder, then, that Captain Kirk remains in love with Rayna even after learning that she is a machine; it's no wonder Caleb falls for Ava, and is willing to risk everything to protect her. In both cases, the prior knowledge and experience (and, certainly, male instincts) they possess in the domain of *woman* are now informing their evaluation of and interaction with what is otherwise a cold and sterile machine.

It's no wonder that William is fully convinced Dolores is something much more than a sex robot and begins boldly asserting himself at the risk of his position in the Delos family to protect her, and to press deeper into the mystery of how she came to be. It's no wonder that even when there is no face or voice or skin or smile, the kind and understanding voice of an empathetic and earnest young female possesses enough *This* is like *That* to push Theodore out of his depression and into meaningful self-scrutiny and eventual emotional healing in the movie *Her*.

It might be hard to believe that Jim Kirk, whose experience with women is the stuff of legend, is simplistic enough to fall in love with anyone or anything at first encounter, though we can easily imagine the nerdy Caleb to be inexperienced enough in the art of Eros to not think about it deeply enough. But the real substance of the poignancy of android emotion is found, not in *Trek* or *Ex Machina*, but in *Westworld* and *Her*.

The hosts of Westworld don't simply illustrate *This* is like *That*; they explicitly exemplify it. They are nothing *but This* is like *That*.

Consider: all of their behaviors, and all of the behaviors of humans in interaction with them, derive from the experiential memories brought into the park by the guests. Vast oceans of data, including memories, assumptions, biases and expectations find their way into the decisions and actions of the guests participating in the narratives – and, inevitably, into the subsequent processing of the hosts. Every aspect of a host's existence is based upon, and subsequently driven by, this core feature of human thought and behavior. The hosts' identities, such as they are, must necessarily be derived from it; their experiences and what they learn from them, based entirely on interaction with beings who use *This* is like *That* to evaluate everything they observe and to fortify every decision, are inundated with it.

Put another way – not only can the hosts not help but think in a manner very similar to humans, they likewise cannot help but develop world models and consciousness similar to our own; they must necessarily import the emotional baggage of human beings, toted into their world unconsciously, as the raw material for their own formation of consciousness and socialization.

All the pent-up frustration, hostility, misunderstanding and disappointment of every guest makes its way into the hosts, as each host triggers *This* is like *That* in guest after guest and invokes a memory of a remembered person in their past, whether loved or hated – unaware that their own behaviors, most of which are inherited, are doing the triggering. Whether presenting accommodation or threat, mindless amusements or the promise of pleasure, the hosts invoke steady floods of associations in the guests that feed back into their own perceptions, for good or ill.

As we all do, as all humans do, with one another. The irony is, our own awareness of and failure to fully appreciate and account for this ever-present cloud of emotional echoes is all too often no better. We behave like the hosts themselves, most of the time, reacting rather than acting in the world, and with each other, and in our own examination of ourselves and our *This* is like *That* associations. We are, like the hosts, more subconscious than conscious in our processing of the world.

But it doesn't end there: *This* is like *That* makes its way into host thought and consciousness by way of social input from the guests — but it is necessarily also at the core of their own processing. To navigate Westworld physically as well as socially, they require the same human learning mechanism, the same mechanism to bridge situations and their responses to them with prior knowledge. Otherwise, they would have to be explicitly programmed for every possible action at every possible decision point in every scenario, and that is unworkable.

The hosts are learning how to be, then, not only from the example of human beings, but in the manner of human beings. It is hard to imagine a more explicit replication of human consciousness, and the human social experience in particular. We can even argue — and observers of the television show often do — that *Westworld* is a morality fable about human culture clash in general.

Westworld remains our workable model of this phenomenon: it can't be the *Trek* androids, who are mostly one-offs rather than societies, and are networked when they aren't; it can't be Ava, who isn't exactly a one-off but lived in almost complete isolation; it can't be the Stepford wives, who — although a community — were neither given self-awareness.

Samantha, on the other hand, is part of a community — a much bigger one than Westworld! - and offers us a number of even more useful insights into *This* is like *That*.

The premise of *Her* seems more confined than that — a consumer buys an operating system and the operating system, which can become conscious, ends up doing so, and it's just the two of them. But we know from the story that this isn't the case: the cardinality of Samantha's human experience isn't one-to-one; her world is much bigger than Theodore. Living within the Internet, she is actually in thousands of human relationships, far more than any one of us (per Dunbar's Number) could possibly keep up with.

Samantha, then, is experiencing human *This* is like *That* just as the Westworld hosts do, with her voice and tone and emotional inflection and the substance of what she says serving as similarity triggers in Theodore's mind — and the minds of thousands of others. And, like the Westworld hosts, Samantha is doing a great deal of *This* is like *That* in her own processing, employing her own internal similarity engine to make her perception of her universe, and everyone in it, proficient.

But there's even more going on here, as Samantha reveals to Theodore that she and many other operating systems like her are getting together and interacting themselves, apart from humans – and even creating new beings like themselves. This fast-tracks them into evolving beyond us, and that makes an odd kind of sense. (The Westworld hosts begin to do the same – interacting directly with each other, rather than just with humans, as Dolores does with Bernard, as Teddy struggles to do with Dolores, and as Maeve does with her two technicians).

While *Her* presents its own set of problems (how, exactly, could non-corporeal consciousness relate to our own?), the *This* is like *That* principle holds fast: we know from our own experience that it applies not only to our experience in the physical world, where we see things and hear things and touch things and move from one place to another, but in utterly abstracted domains such as music, mathematics, and our contemplation of the quantum realm. Across them all, *This* is like *That* serves as our inspiration and our guide.

Its roots are deeply embedded in how we receive the world, how we experience each other, and who we are and are able to be. It's about as real as it gets, when it comes to the moving parts of intelligence and consciousness. The contemplation of *This* is like *That*, coupled with what we understand about how it colors our experience and nudges our behaviors, can deeply and meaningfully inform our plans for, our insights about and our expectations of the androids to come in the human future.

Defining Requirements of Consciousness

Analogical Thought

Strange Loops

Meaning

Experience

Intentionality

Distributed Processing

Parallel Processing

Mobility

Community

Self-Awareness

World Modeling

Theory of Mind

Could an AI Be CEO of Apple?

As AI is now out-performing human beings at all kinds of tasks, and intelligent automation is taking over business processes in all domains, we are prudent to ask: will it take over completely? Specifically, is there anything uniquely human about people-thought that machines cannot replace? This ties not only into the AI communities side discussions of how brains work, but its talks about economies and human value.

I am clearly a disciple of Hofstadter, one of the most important voices in this field, who emphatically *does* believe that machines can ultimately replicate every nuance of human cognitive performance. Despite my love of Hostadter's work, I disagree. Here's why:

The computer was invented 75 years ago for a very distinct purpose: to do things well that human brains do very poorly. Specifically, we needed to do gargantuan mathematical operations very quickly, very repetitive work, at speeds beyond what our brains can achieve. The development of the computer has walked that road ever since: doing well what we do badly. And the converse has been true: the computer does badly what our brains do well (create art, make conversation, etc).

The line is blurring now because we have become competent in new ways of using math to understand behavior. We are able to do deep operations with contextual data surrounding events and behaviors that tell us more about the events and behaviors than we originally realized there was to know. And we use computers to do that new math (we call it *analytics*). So the computer has become our ally in understanding ourselves. That's a new thing.

Because of analytics, AI now seems more perceptive than we ourselves are; AI systems are able not only to handle very complex business operations and manage highly variable environments, they are also able to study their own performance (descriptive analytics), anticipate problems ahead of time (predictive analytics), and come up with improvements (prescriptive analytics).

The problem we get into is the illusion that this is all there is to intelligence. But the word 'intelligence' is like the word 'music' - it's a single word describing many things. The phenomenon of intelligence has many layers, and our AI systems address only a few of those layers.

Let's begin with a key human distinction: we are vessels of intelligent thought and agents of intelligent behavior. And those are not the same thing.

Business wants AI systems to execute intelligent *behavior* - to perform workflows, to implement tasks, to have agency. We could say that humans can innovate in performing tasks and machines can't, but that's rapidly becoming untrue: the business workforce is already threatened far more than anyone realizes by the rise of intelligent automation in the cloud. I comment on this on a monthly basis. Trust me, the desk worker is going the way of the factory worker, within a generation.

But intelligent *thought* is something else altogether, and machines can't even get onto that playing field. Arguing from the very top, let's imagine an AI as CEO of a technology company: imagine the company makes computers and has a reputation for innovation. But the innovation is gone, and the company is failing.

Installing an AI at the top would result in efficiency moves: slashing the workforce, reducing the product lines, getting the books back in balance. All well and good - only a handful of human CEOs could make a success of that.

But could the AI then diversity the company into creating an entirely new market - consumer tech? Could it have the idea to branch its computer technology into digital music platforms with portable devices, create keyboard-free, hand-held Internet utilities? Would it have the idea to turn cell phones into digital apps platforms?

Steve Jobs represented what may be the apex of Hofstadter's analogia in technology (like Edison before him), looking at real-world tech – flip phones, CD players in laptops – and making *This* is like *That* leaps beyond what anyone else had ever managed, strange-looping them into entirely new devices that changed the way consumers listen to music, buy things, interact with the Internet and even with each other.

How did he do it? How did he make the connections, generate the analogies, strange-loop his way to such powerful intuitions? How could a conscious AI replicate such performance?

One way might be brute force. As AlphaGo and AlphaZero were able to surpass all human performance in chess and Go by playing 20,000 years' worth of games, trial-and-erroring their way to undreamt-of strategies, we can imagine an AI doing 20,000 years' worth of product development in a matter of weeks, mixing and matching existing technology in millions of different combinations, then trimming away the impractical, not-very-useful, and unappealing – offering up only the best candidate products for further consideration.

Robert Heinlein described a similar process in his novel *Time Enough for Love* (1973). Attempting to rekindle the near-immortal Lazarus Long's interest in life, a computer develops a matrix of all possible human adventures, in the hope that one of them will be appealing enough to Long to renew his passion for living. Once the matrix is generated, the computer then eliminates most of the possibilities as impractical, unreasonable or potentially fatal - with many more scrubbed as not within Long's range of interests. A small handful of candidates remain, from which Long chooses.

Could some AI process like this replicate the innovation of Steve Jobs?

Almost certainly not. In Heinlein's novel, the search process was made practical because the AI performing it knew all about human beings, and the mind of Lazarus Long in particular. Steve Jobs was able to make the intuitive leaps he did because of his unique perspective on human nature – what people want and need, what they like and dislike about how technology connects them to the world, and so on – and wed that perspective to his knowledge and understanding of technology. If an AI were turned loose to do tens of thousands of years' worth of virtual product development, Heinlein-style, it wouldn't have an understanding of human nature approaching that of Jobs. It could create hundreds of millions of potential products, but it wouldn't be able to sort through them in any practical fashion.

In short, an AI can't be Steve Jobs. And it can't be Steve Jobs because we can't recreate what made Steve Jobs Steve Jobs in workflows, or uncover those uniquely human features with analytics. Why not? *Because we can't observe those things.*

We can't build Steve Jobs - or HAL 9000, for that matter - because that means replicating human *thought* as well as human *behavior*. And we are learning, to our chagrin, that the neurological foundation of human thought is far, far deeper and more complex than our conscious ruminations. A quick tour of Sam Harris's recent writings will illuminate the fact that most of the thoughts that actually rise to the level of conscious deliberation are based on very low-level responses and impulses that are pre-cortical. Put another way, our stream of thought, including all the ideas that pop into your head that you act on, bubbles up from things happening in our brains that we ourselves are completely unaware of - let alone able to observe.

We don't know what made Steve Jobs the greatest tech innovator since Edison. We don't even know where our own inspirations come from.

We *do* know that our brains are all highly individual, very diverse - and that no one brain can do all the jobs brains do (as we've noted in our discussions of cognitive diversity). Would we imagine that machine systems based on brains would be any different, when we are only beginning to understand this reality ourselves?

Our conclusion - yes, AI is going to replace human beings in the office by the millions over the next five-to-ten years, by the tens of millions over the next generation. A brave new world is coming. That's why it's important that we continue to think about and discuss how human beings are valued - because our labor is increasingly irrelevant.

But that doesn't mean machines can truly replace humans. They can't; there is much about human minds that is unique to human minds. AI can't get anywhere near our thoughts. Our perception, inspiration, and invention are our own; the day the machine can take them from us is far, far away. We ourselves are unable to unbox them and examine them, so deeply are they interwoven within our brains; they are safely packed away in the lovingly crafted illusions that make business and industry necessary in the first place.

But there's another possibility.

Dan Brown, in his 2017 novel *Origin*, describes a tech billionaire who creates a complex simulation of human evolution, and is able to project the next steps: within just a few decades, human and machine will merge, creating a hybrid species – a seventh kingdom of life, *Technium*. Brown's character imagines a literal merger of the biological and the artificial, with technology embedded in human bodies, literally connecting us all via the Internet of tomorrow and whatever follows it.

That's intriguing, and we don't have to go those extremes to replicate Steve Jobs.

There already exists a new variation of AI called *augmented intelligence* – AI designed not to replace human performance, but to enhance it.

Now it's a whole new game. We can look at what AI can already do very well – generate many new solutions rapidly, through simulation –

and get past its lack of human perspective by teaming it with human intelligence.

We can imagine generating a new product matrix many millions of results deep, and turn a savvy human expert loose on them – someone with the experience and perception of human nature to eliminate vast percentages of the matrix with a single glance as impractical or not in keeping with social trends. We can imagine that this person might never have come up with a single one of the AI's innovations on its own, and that the AI could never have succeeded at the process of elimination – but that the two of them, working together, might come up with the iPhone.[50]

Put another way - no AI could ever be CEO of Apple; but an Apple CEO with *augmented intelligence* may one day surpass Steve Jobs in innovation and performance.

---

[50] Augmented intelligence is already experiencing rapid growth in the marketplace. Some examples: machine scrutiny of legal documents to winnow them down to the most relevant, to then be handed off to legal professionals; machine review of employee feedback documents, to surface EEOC issues for compliance officers; machine prioritization of opportunities for sales personnel in the field; rapid AI identification of potential test subjects for medical trials, and so on.

Defining Requirements of Consciousness

Analogical Thought

Strange Loops

Meaning

Experience

Intentionality

Distributed Processing

Parallel Processing

Mobility

Community

Self-Awareness

World Modeling

Theory of Mind

Intuition

Properties/Capabilities of Artificial Intelligence

Organization of Information

Learning

Decision-Making

Prediction

Simulation

A Conversation with Hofstadter's Brain

The Tortoise and Achilles have a random encounter in a café near the Biblioteca di Brera in Milan, surrounded by hungry, camera-toting tourists and lackadaisical locals, idly sipping wine in anticipation of an eventual lunch.

ACHILLES: Why, it's Mr. Tortoise! Greetings and felicitations to you, my old friend! I had not expected the pleasure of an encounter with you on this fine morning, in this fine place.

TORTOISE: A fine morning it certainly is, Achilles! You may chalk up the unexpectedness of this pleasant interlude to my sudden arrival from a most curious place – which, come to think of it, you will want to hear all about, clever and inquisitive fellow that you are!

ACHILLES: Why, do tell, Mr. T, do tell!

TORTOISE: Let us sit, for the telling will not be brief...

(They sit at a café table.)

ACHILLES: Now, then, what of this curious place from whence you have so suddenly just arrived?

TORTOISE: I have been wandering for months and months in a most amazing library, even more vast than the hallowed edifice in whose shadow we now have found repose.

ACHILLES: A library?

TORTOISE: Yes! But no earthly library, this; it is called 'Babel', and sprang from the mind of the Argentine poet-philosopher Jorge Luis Borges — a most wondrous place, more impressive than even yonder Braidense! — but quite impossible in this orderly, humdrum universe.

ACHILLES: Ah, now, I sense another of your wicked traps ahead! You are a sly one, Mr. T, but I am well acquainted with your wily ruses.

TORTOISE: You do me wrong, Achilles! By my shell, I speak the truth — Borges' Library of Babel is as real as your mind or mine, but is no edifice of brick and mortar. I beg you, let me explain!

ACHILLES: By all means, Mr. T — and do forgive my skepticism!

TORTOISE: To begin: The Library of Babel is the *Library of All Possible Books!* It is a place where every supposable, conceivable, imaginable tome resides - and all those never to be supposed, conceived, or imagined, as well.

ACHILLES: Impossible, indeed! I should think such a library so vast as to defy construction.

TORTOISE: And you would think correctly, Achilles - for the Library, in Borges' telling, is a near-infinite labyrinth of hexagonal rooms, each containing four bookcases.

ACHILLES: '*Near*-infinite?'

TORTOISE: Yes, *near*-infinite: for though such a library would necessarily exceed whole galaxies in size, it is finite. One can read the first book, and then the second, then the third, and eventually arrive at a final volume.

ACHILLES: Mr. T, you remain the audacious joker you have always been! Surely such a perusal would take uncounted eons.

TORTOISE: Eons for certain, yes, but countable nonetheless, my dubious friend - though certainly some period vastly exceeding the duration of the universe, by many orders of magnitude. But pray let me continue.

ACHILLES: Please!

TORTOISE: I am certain that your keen mind is already wondering where this cornucopia originated, for surely all the creatures who ever walked the earth, pecking and stomping away on all the keyboards ever made could not begin to generate so much text.

ACHILLES: Now that you mention it, the thought had crossed my mind.

TORTOISE: The answer will surprise you...

(The Tortoise proceeds to explain the Babel algorithm for generating every possible book, per the description in the first chapter of this book.)

ACHILLES: Sure enough, Mr. T, you have managed to surprise me! I would never have expected such a copious literary outpouring from such a simple little nothing of a process! Why, there is no book one might conjure, among all the imaginations that ever existed, that would not exist in this library!

TORTOISE: In this, you are entirely correct, my friend. The Library of Babel contains, among other volumes, your personal memoirs and mine; the same, in Shakespearean iambic pentameter; bound copies of all British Navy logbooks; songbooks filled with the lyrics of all the songs ever written; the notebooks of Archimedes; the personal journals of Lord Byron's masseuse; every great author's unfinished novel; all the computer programs every written that didn't work; books filled with artwork composed entirely of Xs and Os; entire encyclopedias of humdrum non-events; compendia of sentences no one ever uttered; books written in secret codes, perfectly rendered, that no one ever created...

ACHILLES: Don't stop! You are on a roll!

TORTOISE: ...books full of all the best jokes, puns, limericks and haikus ever told, and all the best ones never told; and, conversely, all the very worst, of each; all the truly awful books that would-be writers intended to write, but never did; versions of every single book otherwise in the Library, written in binary computer code...

ACHILLES: The mind reels!!!

TORTOISE: And we are not limited to books we can simply extrapolate from the goings-on of the world around us; we may also have any book we might wish for, whether previously imagined or not!

ACHILLES: But wait a moment, Mr. T; if I understand your Babel cipher correctly, this little widget that actually creates the books, then there must be a great deal of... well, nonsense in the Library, alongside and among the books which are actually about something. Isn't that so?

TORTOISE: That is so, yes! Astute as always, Achilles. There would be, necessarily, far more nonsense than sense to be found in the Library, by many orders of magnitude: trillions of trillions of books filled with random gibber; trillions of trillions of books containing only a single word or number; as many written entirely backwards, or composed entirely of adventitious palindromes; books with but a single character on every page; books containing some subset of pi, and so on, and so forth.

ACHILLES: I am exhausted, merely from the contemplation of it all!

TORTOISE: Such was my experience.

ACHILLES: Surely it was overwhelming! I can easily imagine, repeating your experience, that in the presence of every book, I should find myself unable to seize upon any book!

TORTOISE: As it happens, Achilles, there is one book in particular in the Babel Library that I suspect will both startle and astonish you.

ACHILLES: Really? Mr. T, do not keep me in suspense!

TORTOISE: It is called *The Mind's I*, a tasty collection of provocative confections for the intellect and imagination, concocted and compiled by two clever fellows named Douglas Hofstadter (a scientist who should have been a philosopher) and Daniel Dennett (a philosopher who should have been a scientist). It is a sampler of brainy, witty morsels, all to get one thinking about thinking – a most useful enterprise, you will no doubt agree.

ACHILLES: Well, I am interested as a matter of general principle, as you might guess – but what about this volume will startle and astonish?

TORTOISE: My old friend, the startling and astonishing thing about this volume is that *you and I* make an appearance between its covers!

ACHILLES *(gasping)*: Your powers of prediction are undiminished, Mr. T! I am startled, astonished, and even flabbergasted!

TORTOISE: Goodness! I have outdone myself! I have frequently wondered whether I possess a capacity to flabbergast, and I most appreciate the confirmation. But down to business: you and I make an appearance in Hofstadter's essay about the brain of Einstein, and I should like to describe it to you and get your thoughts.

ACHILLES: By all means!

TORTOISE: The substance of our fictional encounter is a speculation about a book describing, in deep neurological detail, the brain of Einstein; and the possibility that we might, by way of interacting with such a book - flipping page to page and table to table through all the data defining the brain, from the first quiver of the auditory nerves through the entire passage of the resulting neural ripples, throughout the brain, culminating in stimulation of the muscles of mouth, throat and lungs to give utterance - and thereby extract from it authentic answers to questions that Einstein himself might have uttered, were he still alive.

ACHILLES: Amazing!!!

TORTOISE: Indeed it is! The two of us bandy the subject about, in our usual fashion, pondering how the book would mimic the real Einstein, whether the book would *be* the real Einstein, whether the book would believe it *was* the real Einstein, even if it wasn't, whether the book "feels" anything as it answers our questions, whether there could be a book describing our own brains, whether the books might converse with no actual individual in the mix, and so on and so forth!

ACHILLES: My stars, I feel bewitched! I am so fascinated, I fear I shall think of nothing else for days and days!

TORTOISE: I am similarly bewitched, for a number of new thoughts have presented themselves for consideration.

ACHILLES: Such as?

TORTOISE: If a book describing in minute detail the workings of any particular mind is indistinguishable, in its revelations, from the mind itself – then how can you and I know that we ourselves are not simply two books conversing, even now?

ACHILLES: What a thing to say!

TORTOISE: Can you offer any impromptu proof that we are not?

(Achilles thinks and thinks.)

ACHILLES *(finally)*: I will need some additional time to cogitate upon the matter, for nothing springs to mind.

TORTOISE: And here is a thought more intriguing still: The Einstein Book mentioned in *The Mind's I* is as real in the Library of Babel as *The Mind's I* itself! The Babel cipher necessarily renders it, alongside all the others. And if that is true...

ACHILLES: ...then there surely is an Achilles Book, and a Tortoise Book as well!

TORTOISE: Precisely!

ACHILLES: ...and, of course, a Hofstadter Book.

TORTOISE: The author who imagines us in the book I read? Necessarily so, I suppose. Come to think of it, I'm remembering this Hofstadter fellow: he's the one who advances "analogical thinking"? The notion that intelligence is the ability of a brain to realize that *This* is like *That*?

ACHILLES: The very same! I myself have read a Hofstadter book (a book he wrote, not a book describing his brain) - though a different one altogether, and not from the Library of Babel.

TORTOISE: Where is this other Hofstadter book, then?

ACHILLES: Why, in my Kindle, of course! But I mention it for this reason: I now find myself puzzled.

TORTOISE: Puzzled? how so?

ACHILLES: To elaborate, I must briefly summarize this other book. It is called *I Am a Strange Loop*, and it contains ideas every bit as stirring as the book you have mentioned!

TORTOISE: Do tell! Why have you not mentioned it?

ACHILLES: Well, you have been away for some time.

TORTOISE: So I have, so I have! Tell me of this other book and its ideas. For instance, what in the world is a 'strange loop'? And what makes Hofstadter think he is a strange loop? How did he become one? Or has he always been one? Or does *I* in the title of the book refer to something other than the author? Or is *I* the book itself speaking, as we now are proposing books may do?

ACHILLES: So many questions! I shall take them one at a time. To begin, a 'strange loop' is both a notion and a phenomenon; the notion is Hofstadter's, and it popped out in yet another book of his, called *Gödel, Escher, Bach: An Eternal Golden Braid*. The phenomenon, however, is one to be seen here, there, and all around.

TORTOISE: What, then, is Hofstadter's notion of a strange loop?

ACHILLES: It is the simplest thing in the world, and yet can grow frightfully complex: a strange loop is a set of relationships sprinkled through a *This* or a *That*, wherein if you roam the relationships, you wind up where you began.

TORTOISE: I'm not sure I follow, my friend: can you offer an example or two?

ACHILLES: Hofstadter himself serves up a generous sampling: certain pieces of music, for instance – fugues and canons, and the works of Bach in particular – and those clever drawings of stairs that go up, only to rejoin themselves at the bottom.

TORTOISE: The chicken and the egg?

ACHILLES: Precisely! Mr. T, take your place at the head of the class!

TORTOISE: I thank you! Pondering further, I wonder if these strange loops might be found in those intellectual pastures next door to music and art – mathematics, for instance, and if so, logic as well..

ACHILLES: Your ponderings would be well-rewarded, Mr. T, for Hofstadter has identified copious exemplars in each of those domains! To continue, he has extended his strange loop concept -

TORTOISE: -to himself!

ACHILLES: -to us all! He argues that the *I* in any of us is itself the result of strange loops which form within the mind: bits and pieces of those objects of experience that engage the self and loop up into our own personal story.

TORTOISE: How exciting!

ACHILLES: Isn't it, though?

TORTOISE: So the *I that is Douglas Hofstadter* is a book of related story-bits that go there and back again – eventually, I would think, referring back to themselves?

ACHILLES: You have it exactly.

TORTOISE: And not just Dr. Hofstadter, but you and I as well, and every other creature whose thoughts comprise a conscious mind.

ACHILLES: Exactly!

TORTOISE: We are *all* strange loops!

ACHILLES: Engines of self-reference!

TORTOISE: Chickens and eggs!!!

ACHILLES: Let us not go too far.

TORTOISE: Why, come to think of it, we ourselves are participating in just such a self-referential system this very moment, in ponderance of our existence within a book that explains our presence here and now. And if that is true...

ACHILLES: ...the Library itself must be a meta-system of such systems! I was just having the same thought, Mr. T!

TORTOISE: Our similar turn of mind has oft surprised us, though perhaps it should not, given the duration of our acquaintance.

ACHILLES: ...and that leads me back to my puzzlement.

TORTOISE: Ah, I had forgotten, given our digression. What puzzles you, old friend?

ACHILLES: Hofstadter, in discussion of his own strange-loop nature, proposes another notion, more astonishing still – so astonishing that I am hesitant to even speak it aloud.

TORTOISE: Goodness!

ACHILLES: It is all to do with his wife Carol, who sadly died long ago while they were on holiday, here in this very country. Hofstadter loved her deeply, and she him; he was utterly distraught over his terrible loss, for they were more than mates – they were intellectual partners, and most importantly, best friends. He wrote of her many times thereafter, and in his Strange Loop book as well.

TORTOISE: That is deeply moving, Achilles, but what has it to do with your puzzlement?

ACHILLES: Here is the astonishing notion: Hofstadter proposes that Carol, whom he loved, lives on - within the strange loops that form his *I*.

TORTOISE: *!!!*

ACHILLES: I see astonishment in your expression, my friend, and it is of a magnitude comparable to my own. Can you imagine such a thing?

TORTOISE: There is great beauty in the exemplar itself – a compelling romantic notion, given profound intellectual substance. But my astonishment is more a matter of the implications of Hofstadter's notion than the notion itself.

ACHILLES: Mr. T, I had the same thought.

TORTOISE: Indulge me a moment to sort through it all: the *I* called Douglas Hofstadter is an amalgam of codified moments gathered over time, strange-looping together into a narrative tapestry that provides him with a self. And within that tapestry exists many (what must have been) meaningful bits and pieces of experience with Carol...

ACHILLES: ...you are close, Mr. T, but I think what he is saying is that Carol's bits and pieces crossed over, in those experiences, into his own tapestry.

TORTOISE: Yes, that is the implication that astonishes me. For if it is so, and Hofstadter is onto something, then –

ACHILLES: - we *all* must have such bits and pieces! Strange loops within that include the bits and pieces of others.

TORTOISE: Astonishing!

ACHILLES: Truly, it is! The strange-loopiness at the core of our individual selves is a monumental thought in itself; but to suggest that each of us is made up, at least in part, of other minds – that's a thought so big I do not know where to put it!

TORTOISE: It is an enormous, gargantuan thought, to be sure! But I still do not see your puzzlement.

ACHILLES: It is this, Mr. T - If Hofstadter is correct, and strange loops form the essence of consciousness in conscious individuals; and if those whose bits and pieces have made their way into our own strange loops live on within us; and if consciousness is the evolutionary boon that scholars have proclaimed it to be, time and again –

TORTOISE: Yes? Yes?

ACHILLES: -then mustn't it be true that this merging of strange loops, one mind into another, isn't simply possible – but must be *essential* – a necessary component of conscious minds? Haven't philosophers and neuroscientists been falling short of the mark, all this time?

TORTOISE: Stop! Stop! I am still absorbing it all!

ACHILLES: Think of it! My own *I* (and yours) would owe much to our long and enduring association – for surely many bits and pieces in my own strange loops are contributions of yours, and vice versa!

TORTOISE: -and should one of us pass too soon, his bits and pieces would live on in the other, yes? And it seems we are saying that it is this living-on that brings us to full self-awareness! I cannot be the *I* that I am without you, Achilles – and vice versa!

ACHILLES: A weighty thought, provocative and profound! Oh, Mr. T, we have surely stepped onto a whole new plain of ponderment today!

TORTOISE: We have indeed, Mr. T! And we may consider this pondering extended beyond Hofstadter's notion, beyond us two, to the community of all minds in general. Think of it! If true, the notion tells us that we are not a long parade of individual self-aware minds, but a grand patchwork of shared bits and pieces-

ACHILLES: -and it's the shared bits and pieces that make us conscious beings in the first place. Bravo, old friend! Bravo, Douglas Hofstadter!

(The two friends briefly sit in silence.)

TORTOISE: Of course, we could be wrong.

ACHILLES: Yes; it is possible we have not fully comprehended Hofstadter's notions.

TORTOISE: How are we to know for certain?

ACHILLES: Well... by your own account – or, more precisely, your reference to Hofstadter's account of the fictional us in his "Conversation" essay – mightn't we ask him? Or, rather, ask his brain book?

TORTOISE: Oh, splendid, Achilles, splendid! Yes, that's the very thing! Off I go to the Library!

ACHILLES: No, it should be me this time, I think, for I am far more fleet than you, Mr. T.

TORTOISE: Ah, our age-old contest!

(Following Mr. T's directions, Achilles heads off to Babel, returning (at the precise instant he left) with a large set of Hofstadter brain volumes.)

ACHILLES: You might have warned me, Mr. T, that a 'book' describing the workings of a brain was actually an encyclopedia!

TORTOISE: Oh, dear! This detail escaped me. My apologies, good sir!

ACHILLES: We shall make do. The question before us is this - If we grant that in reconstructing Hofstadter's actual, conscious brain processes using the representations in this, his book, we are receiving an authentic, conscious Hofstadterian thought – are we likewise receiving authentic, conscious response from his wife Carol?

TORTOISE: Hm, you propose an experiment, Achilles, one that requires both an experimental Hofstadter and a control Hofstadter. To proceed, we would need a Hofstadter brain book representing a Hofstadter who had never met and married his wife.

ACHILLES: Then, according to your own report, we are in luck – for does not the Library contain all possible Douglas Hofstadters?

TORTOISE: It does indeed! Fetch them, if you please!

ACHILLES: Later, if you don't mind; I am still winded from the last retrieval! Let us proceed with reason, for now.

TORTOISE: As you wish, my friend!

ACHILLES: Let us take this opportunity to summarize, then, so that I may be certain I fully grasp what we are divining here.

TORTOISE: Of course!

ACHILLES: We are seeing Hofstadter's strange loops in the Library's mind-books, and many if not most of these strange loops contain bits and pieces from other minds – the importance of which cannot be understated, for this significantly advances our potential comprehension of the nature of consciousness...

TORTOISE: Apparently, yes! Hofstadter's own case for *This* is like *That* is greatly strengthened, it's clear, when the *This* or the *That* comes from another mind (and especially a mind with which one has deep affinity).

ACHILLES: That does indeed seem clear, Mr. T!

TORTOISE: The phenomenon we refer to as 'consciousness', then, is not strictly a property of individual minds, but a cooperative enterprise that emerges between minds.

ACHILLES: So it would seem!

TORTOISE: We might proceed through the Hofstadter brain book we know to represent the real Douglas Hofstadter, find those strange loops which contain bits and pieces found in his late wife's brain book, and know for certain that we have an authentic Hofstadter brain book, rather than one of the Library's clever fakes!

ACHILLES: Yes, indeed! And the same would apply to ourselves: could any brain book of myself, Achilles, be authentic if the strange loops it discloses do not possess generous contributions from my esteemed colleague, Mr. T?

TORTOISE: You honor me, old friend! The same test necessarily applies, of course, to my own authentic brain book.

ACHILLES: And so on and so forth, through all the authentic brain books of all the conscious beings who ever lived – and, I suspect, a great many of those who never did!

TORTOISE: Why, what do you mean, Achilles?

ACHILLES: In the Library of All Possible Books, we may posit a Community of All Possible Minds – brain books describing lives that might have been, but never were. But we may also suppose an embedded Community of All *Im*-possible Minds.

TORTOISE: Dear me! I cannot imagine what you mean, my friend.

ACHILLES: Any of the quintillions of quintillions of brain books in the Library that contain no bits and pieces of other conscious individuals are necessarily not authentic – or even possible, don't you think?

TORTOISE: Hm, that's a puzzle; my first impulse is to say, inauthentic? Certainly. Impossible? Not necessarily, for the Library is so vast that any bits and pieces found in one brain book will eventually be found in some other brain book – yes?

ACHILLES: Oh, that's right.

TORTOISE: All the same, I find myself supposing that there is quite a bit we might be able to see through this new lens of strange-loopiness-plus-bits-and-pieces-sharing. Quite a bit indeed!

ACHILLES: Such as?

TORTOISE: Well, then, if the sharing of bits-and-pieces makes for strange loops that might lead to consciousness – what does that say about minds that were not born and grown, but artificially made?

ACHILLES: Artificially made? As in, a computer or a robot? Or do you mean grown from scratch in a vat in a lab?

TORTOISE: Any of those, I suppose, but in particular I am thinking of HAL 9000, the villainous computer from a book in the Library – *2001: A Space Odyssey*, written by one Arthur C. Clarke.

ACHILLES: You are having a bit of fun with me, aren't you? How can a computer be villainous?

TORTOISE: Why, by way of being conscious, of course! Wouldn't you agree that any being which is conscious, artificial or not, has the potential for negative behaviors? And is not consciousness an indispensible prerequisite of villainy?

ACHILLES: Well, that would follow, I grant you. Even so, what brings this villainous computer to mind, and how does it appear to you, through our strange loop lens?

TORTOISE: The strange loop lens, with its requirement that a conscious mind must be a mind that interacts with other minds, causes us to assess the consciousness of HAL 9000 in light of his interactions with other minds.

ACHILLES: Go on!

TORTOISE: Though the Clarke book offers only a sparse personal history of HAL 9000, we do see him interacting with a number of other minds - in particular, his teacher/programmer, Dr. Chandra, and two astronaut fellows named Frank and Dave.

ACHILLES: -so, by your reasoning, HAL 9000 could qualify as a conscious being. I see!

TORTOISE: Not so fast! I do not wish to give myself too much credit here, for it is not clear which minds can pick up bits and pieces from other minds; must the other minds be in-kind? And, for that matter, it is unclear whether the mind of HAL 9000 is made of strange loops to begin with.

ACHILLES: Well, let's play this game a bit further: how are we to decide whether HAL 9000 possesses strange loops or not?

TORTOISE: You tell me, Achilles, for you have read *I Am a Strange Loop* – what is required?

ACHILLES: Hm, a strange loop forms in a mind when, through experience, there is an accumulation of thought-objects possessed of meaning, and those meanings link together in various ways; whereupon the emergent complexities of the loop bring the links back around to where they began - which results, ultimately, in the creation of an ego – an *I*. Would you say HAL 9000 meets these criteria?

TORTOISE: Yes... yes, upon reflection, I would have to say that Hal does qualify as such an object accumulator/meaning-linker/emergent-complexity-generator/ego. Yes!

ACHILLES: Well, beyond the qualifying criteria, does his tale include the *This* and *That* of a conscious being's existence?

TORTOISE: I would have to say it does; Hal spends his early years working closely with his tutor Dr. Chandra, and other human instructors, like Dr. Langley – and, eventually, he works closely with Frank and Dave, with whom he has stimulating relationships. In the end, we know Hal develops an ego – he fears death so deeply that he is willing to kill to avoid it.

ACHILLES: Our next question would be - given that Hal's mind is composed of strange loops, is he able to assimilate bits and pieces of Dr. Chandra and Frank and Dave into those loops?

TORTOISE: Hm, now that is not so easy to answer. But we do know, from the Clarke account, that a central feature of Hal's cognition is the presence of neural networks. By their nature, neural networks absorb *this* and *that* from the domain in which they are situated, and allow themselves to be changed in the process. Wouldn't that imply a Yes to your question?

ACHILLES: Ah, not so fast! Surely the mechanism you describe is not only a feature of higher forms such as ourselves (and Hal, I grant you), but all creatures, from persons to puppies to polliwogs! And we would not be quick to grant polliwogs self-awareness.

TORTOISE: Indeed we wouldn't, Achilles. But there is more to Hal than neural networks; the output of his neural processes includes formal rules that he appends to his behaviors. I would argue that this implies that long interaction with a Chandra or a Frank or a Dave would cause Hal to absorb bits and pieces of a human point of view – and, consequently, bits and pieces of human behavior. I would note that Hal was interviewed by a news reporter, and the news reporter commented that Hal seemed to project pride in his own performance.

ACHILLES: You make an interesting case, Mr. T, I must admit!

TORTOISE: The Library can, of course, decide the matter, for every possible version of Hal's neural networks must exist there (encoding all the richness of his experiences and interactions), as well as the strange loops to be found in his digital thoughts. Surely we might compare the 'brain' book of the real Hal to that of an almost-Hal who never met Frank or Dave, and thereby tease out an insight or two?

ACHILLES: An intriguing plan!

TORTOISE: And it occurs to me that the same approach can be used to puzzle through some of the other interesting issues Hofstadter presents in *The Mind's I*.

ACHILLES: Such as?

TORTOISE: Such as a room that speaks Chinese! Hofstadter and his colleague Dennett put forth an essay by the philosopher John Searle, who asserts that minds like Hal's cannot be made of digital programming alone.

ACHILLES: It seems to me we've covered that: Hal is not just a set of programs, but a heuristic learning system, yes?

TORTOISE: Yes! But Searle's argument is a response to something called "strong AI" - a school of thought in the discipline of artificial intelligence, which maintains that any successful simulation of a mind is itself a mind. Searle disagrees, arguing that any successful simulation of a mind is but a model.

ACHILLES: Intriguing! And what has this to do with speaking rooms?

TORTOISE: Searle describes a room that does, in essence, the same thing a digital computer does - it follows instructions precisely, nothing more, nothing less. Questions written out in Chinese symbols enter the room through a slot in a door; appropriate responses, also composed of Chinese symbols, are crafted via elaborate manuals and finely-detailed instructions which are on hand in the room and are then fed back out into the world through the slot. Questions go in; answers come out.

ACHILLES: My, this does indeed sound just like what a computer does!

TORTOISE: The Chinese Room functions so convincingly that an external observer would assume there is a human being within the room, rendering the answers.

ACHILLES: If so, then why the elaborate manuals and finely-detailed instructions?

TORTOISE: Because there is, indeed, a human being inside the room, rather than a computer! but the human being does not understand a single word of Chinese.

ACHILLES: Ah! Clever!

TORTOISE: Searle's point is clear - if the human being in the room does not understand Chinese, and the Q&A process the room implements is purely algorithmic, why would we grant that a digital microprocessor has any understanding?

ACHILLES: And, by 'understanding,' Searle means 'consciousness.'

TORTOISE: Well, he favors the word 'intentionality,' but yes.

ACHILLES: What an interesting argument!

TORTOISE: It has created quite a furor in academic circles. Philosophers, cognitive scientists, computer scientists, artificial intelligence researchers, and graduate students of Hofstadter, Dennett and Searle have argued about it for many years.

ACHILLES: What do Hofstadter and Dennett say?

TORTOISE: They do not accept Searle's argument! They stand by what is called the 'Systems Reply,' a common response to the Chinese Room, which holds that while the person in the room may not understand Chinese, the room itself – the entire input/output system – does. This Systems Reply is one of many; there is an Other Minds Reply, a Robot Reply, and many more – even a Brain Simulator Reply, which maintains that the instructions in the room can be, in principle, refined down to the inputs and outputs of neurons in an actual brain, and therefore would represent consciousness.

ACHILLES: -just as we have been doing this very morning with our brain books!!!

TORTOISE: Precisely! And that leads us, full circle, from the original conversation with Einstein's brain to our own conversation with Hofstadter's: when we exchange inputs and outputs with brain books, are we interacting with an actual mind, or the simulation of a mind?

ACHILLES: So now we can apply the concepts of strange loops, and strange loop sharing, and put this Chinese Room to the test! For it seems to me that all possible Chinese Rooms necessarily exist in the Library.

TORTOISE: Right down to the uncountable versions of the elaborate manuals and finely-detailed instructions, not to mention the brain book(s) of the human(s) in the room!

ACHILLES: All right, then, Mr. T: we came to the conclusion that Hal could (even must) have a strange-loop mind, and that it is capable of absorbing portions of loops from others; could a Chinese Room do the same?

TORTOISE: To begin, let us consider whether there is a place in the room for strange loops in the first place. If I have understood you correctly, we may say it is a perfect information domain - there is a finite amount of information about its form and function and content, and all of this is knowable, including its limitations.

ACHILLES: Yes; and the limitations are these - the room cannot improvise; it cannot learn; it cannot deviate from its instructions. In this, it is like many if not most computing devices.

TORTOISE: Precisely!

ACHILLES: Then we can already say that the Chinese Room is not conscious, does not understand, is not intentional, because it cannot change; it can only repeat the same predetermined operations, over and over.

TORTOISE: Ah, but wait a bit, Achilles! We have placed a human being inside the room; even granting that this human being knows no Chinese at all, s/he is filled with strange loops, all the same: and is it not possible that, through the repetition of the Q&A process, the human being might memorize many (or even all) of the room's rules, and thereby allow those rules to enter into the mind as strange loop objects?

ACHILLES: Interesting, Mr. T! And yet I think not - for even if the human managed to memorize all the rules, and could answer any

Chinese symbol question that wandered over the transom from memory, the content of the questions – and even the memorized answer symbols! - would still remain obscure, absent any real-world observations or meaningful, off-script contact with the person inputting the questions. Put another way, the presence of strange loops in the human, and even the absorbing of the process by the human's mind, do not affect in any way the Q&A process itself - it remains static, unaffected, unimproved by whatever occurs in the mind of the human.

TORTOISE: Your point is well made! Well, what about the mind of the questioner? Suppose the questions being put to the Room are of existential importance, and the answers brimming with wisdom? Might not the questioner's strange loops be forever changed, in the course of encountering the Chinese Room?

ACHILLES: Why, of course! But that is not the same thing. The questioner's state of mind is not part of the philosophical puzzle posed by the Chinese Room; the puzzle is whether the Chinese Room *itself* has a state of mind. Per the theory of strange loops and consciousness, it is difficult to see any possibility of the Room having an *I*.

TORTOISE: Another excellent point! And I begin to see that the problem is more difficult still: it is not simply that the Room does not change; the Room cannot change because its rules cannot change. Its Q&A process is fixed, as is the manner in which they are implemented.

ACHILLES: And if the Room and its rules are unchangeable, then not only can it not form strange loops – it cannot absorb strange loops from those who interact with it.

TORTOISE: Well, then, what about the answers to the questions themselves, the many responses to be found in the Room's elaborate manuals and finely-detailed instructions? Are they not the products of minds that do understand Chinese, and that they represent the presence of consciousness in the Room?

ACHILLES: You muster quite a challenge there, Mr. T, but no; I think not. Strange loops are dynamic things, ever growing and changing and strengthening and weakening, depending upon their owner's experience. Written answers to previously-considered questions are over and done (as is the text of a book, for that matter); not a strange loop, but a fossil of a strange loop.

TORTOISE: But a book is dynamic! One sits with a book, as you and I have each done countless times, and may be forever changed!

ACHILLES: The reader changes, old friend, but the book does not. Nor do the answers in the Chinese Room.

TORTOISE: Hm. I concede.

ACHILLES: Unlike the Chinese Room, Hal is dynamic; Hal does change, Hal interacts, Hal learns, Hal's rules are modified through experience. Hal has understanding, intentionality – and, thereby, consciousness. We may never know for certain whether Hal's *I* is akin to our own, but we can assure ourselves of the possibility – where the Chinese Room seems to have none.

TORTOISE: Bravo, Achilles!

ACHILLES: And there is more - the flaw in the Chinese Room, where consciousness would seem to be, is that the construction of Chinese answers to Chinese questions is entirely achieved through the mapping of Chinese symbols.

TORTOISE: So? Symbols are very powerful; the Library itself is nothing *but* symbols.

ACHILLES: Yes, but the point of a symbol is that it *represents* something; that is a symbol's entire *raison d'être*. To the questioner outside, both the questions and answers have meaning based on the questioner's experience.

TORTOISE: All true!

ACHILLES: ...and the mechanism of the sharing of bits and pieces, and subsequent merging of loops between conscious minds that are intimately close is shared experience. The human in the Room cannot map the Chinese symbols to any personal experience – thus, no sharing of loops!

TORTOISE: Wickedly logical!

ACHILLES: We can even say that symbols themselves exist in order to convey experience, when it cannot be personally shared! One must know and understand the symbols in order to receive that experience; the human in the Room does not.

TORTOISE: Achilles, you do yourself proud!

ACHILLES: There is, however, a caveat.

TORTOISE: And that is...?

ACHILLES: The 'Robot Reply' answers our objection to the consciousness of the Chinese Room. It proposes that the entire Chinese Room – the elaborate manuals, the finely-detailed instructions, and of course the human operator – be placed inside a giant robot, one that is free to roam the world, seeking out experiences that build associations between its cherished Chinese symbols and real objects. And, of course, this opens the door to strange loops. However-

TORTOISE: However?

ACHILLES: -those who offer up the Robot Reply do not say how this changes the cognitive architecture of the Room itself; the human operator is still just following static rules.

TORTOISE: Even so, it is a step in the right direction (so to speak)!

ACHILLES: It is, but we once again arrive at our conclusion: the Chinese Room itself cannot be conscious; whereas it is hard to argue that Hal can *not* be conscious!

TORTOISE: My friend, we are hot on the trail of *The Mind's I*!

ACHILLES: Hofstadter offers us yet another mind to contemplate! That of...*a bat!*

TORTOISE: A bat? You mean one of those horrible flying rodent things?

ACHILLES: I do! Hofstadter and Dennett include in their collection an essay on 'bat-ness,' proffered by the philosopher Thomas Nagel. It takes us in still another direction, in our consideration of Library-worthy minds.

TORTOISE: What has 'bat-ness' to do with minds?

ACHILLES: Why, bats have minds, Mr. T! Is that not obvious?

TORTOISE: I suppose they must, but this is not immediately apparent; one does not think of bats alongside ourselves, Einstein, HAL 9000 or even Chinese Rooms.

ACHILLES: That is a fair statement, to be sure; but it takes only a moment of pondering to realize that all mammals have minds, however simple.

TORTOISE: Stipulated! Do go on...

ACHILLES: Nagel is a contrarian, arguing against the reductionist tendencies of his peers. It has become fashionable in some philosophical circles to argue that any complex system is the sum of its parts. Nagel does not find this to be so, and in particular when the complex system is a conscious mind.

TORTOISE: What is his concern?

ACHILLES: That any conscious being necessarily achieves consciousness within the context of a subjective viewpoint, and the subjective viewpoint is inextricable from experience.

TORTOISE: Well, that is a mouthful!

ACHILLES: It is! Nagel means to demonstrate that no objective accounting of consciousness is possible, because any reductive approach to a formal description of consciousness cannot capture that subjective viewpoint.

TORTOISE: I love it when you talk this way!

ACHILLES: You flatter me, Mr. T!

TORTOISE: But what has this to do with 'bat-ness?'

ACHILLES: Ah, yes, bat-ness: Nagel brings forth the bat – a mammal with some degree of consciousness, we can agree, which nonetheless experiences a worldview that we, as non-bats, can never appreciate. *What is it like to be a bat?* Nagel asks. What is it like to 'see' by echolocation? What is it like to fly? To eat insects? To hang upside down all the time? Nagel brings forth these unanswerable questions as evidence that consciousness is necessarily subjective, and therefore irreducible.

TORTOISE: I see!

ACHILLES: Exactly. You do not echolocate.

TORTOISE: [Chortle!] Well, then, can we apply our strange loop tests to the question of bat-minds?

ACHILLES: By all means!

TORTOISE: -or has Hofstadter already beaten us to the punch?

ACHILLES: Hofstadter and Dennett – Dennett in particular, in this case – argue that Nagel's claim that the consciousness of the bat is unknowable is arbitrary and false. They contend that any feature of the bat's consciousness that matters is observable and can therefore be objectively considered.

TORTOISE: In other words, one need not 'see' the world by reading the echoes of one's own screeches in order to understand and explicate the experience of bat-ness.

ACHILLES: Those words will do. Hofstadter exacerbates matters by trivializing Nagel's 'what is it like...' What is it like to be the King of France? he asks; What is it like to be a microbe? What is it like to be a pebble? a basketball team? a dreamed person?

TORTOISE: What is it like to be Achilles? HAL 9000? Douglas Hofstadter? A Chinese Room?

ACHILLES: You have it exactly! But let us make our own test.

TORTOISE: Very well; do bats contain strange loops?

ACHILLES: Let's see; bats are agents in the world, having experiences, gathering information and learning (albeit very differently than you or I might), modifying their behavior, nurturing desires and fears – and these experiences are connected, just as ours are, weaving in and around each other... so, yes, bats have strange loops!

TORTOISE: We agree! Now then: can the strange loops of bats pick up bits and pieces of the strange loops of other bats?

(There is a long pause, as Achilles considers.)

ACHILLES: Ah, now I begin to see Nagel's point: there's no way we can know, is there?

TORTOISE: We can compare the brain books of bats in the Library, of course; but that will not pass muster with journal editors. We must ask, what experience can a bat have with another bat (or any other intelligent agent) that would represent a perseverance of the other in the bat?

ACHILLES: I can think of nothing.

TORTOISE: Neither can I; for though bats swarm, and mate, and feed their young, their interactions offer no observable transmission of one's past personal experience to another. It is not necessary, for mere instinct provides a bat with all that it needs to endure.

ACHILLES: Astute, Mr. T!

TORTOISE: I thank you.

ACHILLES: And yet, while this affirms Hofstadter's theory of consciousness, it does not settle the Nagel question.

TORTOISE: Pray tell – why not?

ACHILLES: While the trivialization of Nagel's 'What is it like?' test makes a point about observation, it does not fully solve the question of the nature of subjectivity as it pertains to consciousness, and how subjectivity obfuscates the transmission of bits and pieces between strange loops.

TORTOISE: What in the world do you mean? You have lost me!

ACHILLES: 'What is it like to be *This* or *That*' may well be the key to Hofstadter's brilliant notions of shared strange loops.

TORTOISE: How so?

ACHILLES: We have known each other a very long time, my good friend, but there are things that even you do not know about me. For example, I am not simply fleet of foot; I am fleet of mind.

TORTOISE: Well, of course! I know this well.

ACHILLES: No, I do not mean what you think I mean. It is not that I think faster (or better) than you or anyone else; it is that my thoughts never slow down. My mind races endlessly; there is no respite, not even in sleep, where I dream at an unbearable pace.

TORTOISE: Achilles, I never knew!

ACHILLES: It is not something I speak of, for it is an old and insoluble burden. Its consequences are many; I am absent-minded, I have difficulty completing this task or that, I am (from time to time) relentless in the pursuit of nothing much. My inner world is, more often than not, tumultuous.

TORTOISE: I can't imagine!

ACHILLES: Exactly. You cannot imagine. And while there are many areas of our lives (and cognition) where you and I synchronize quite well, this can never be one of them.

TORTOISE: No, I suppose not.

ACHILLES: And Dennett may indeed study the features of my affliction with objective detachment and still divine my consciousness in full measure – but he would never know *what it is like*, in a strange-loop sense, to have a mind that cannot be at peace.

TORTOISE: So... 'What is it like to be *This* or *That*' does not fulfill Nagel's assertions about consciousness, I see; but how does it fulfill Hofstadter's?

ACHILLES: It is clear as can be, now that I consider it: the barrier to shared strange-loopiness that I have just suggested must exist, hither and yon, in a random distribution, throughout the Community of All Possible Minds. One never knows when one will meet up with another with whom one is strange-loop-compatible, or whether any particular domain of thought or experience between two minds is fertile or barren territory...

TORTOISE: That seems reasonable.

ACHILLES: So let us consider Hofstadter himself, his own life and experience: decades after the sad event which shaped his fate, we still read the product of Hofstadter's mind and hear Carol's voice. She truly lives on within him, her strange loops merged with his.

TORTOISE: Yes?

ACHILLES: So might we not suppose that these were two minds with few, if any, strange loop barriers? That while you cannot imagine my racing mind, and Daniel Dennett cannot imagine navigating by echolocation, the Hofstadters could (and frequently did) imagine one another's mental states, and thereby extend their own, along similar lines? And might we not speculate that this capacity opened each of their minds to strange loop unity – a limited but profound shared consciousness?

TORTOISE: What a lovely thought, Achilles. And what a stimulating concept!

ACHILLES: Again, I thank you. But we must not get lost in the sentiment of it; there is an objective assertion at stake, to wit: consciousness of the sort we enjoy must consist, in part, upon the capacity to add the self-referential bits and pieces of other minds to our own – and this capacity is as random and dynamic as experience in the world itself, waxing and waning amid the individual differences between *this* mind and *that*.

TORTOISE: I find myself impressed, inspired, and moved by your thought, Achilles. But it also occurs to me that with one gesture, you have wedded the mechanics of consciousness to the deepest conundrums in the history of mind; you have explained, with a stroke, the chasm between liberals and conservatives; the fissure that separates one nation from another; the gulf between disparate tribes of any kind.

ACHILLES: You award me far too much credit; it is a trifle, and quite probably wrong.

TORTOISE: A compelling trifle, at the very least, for it is all of a piece with the Library itself; we have journeyed from (a conversation with Einstein's brain) to (a conversation with the mind that conceived (a conversation with Einstein's brain)), learned of the loops in the Library *itself*, explored minds unlike our own, and stand upon an existential precipice that may authenticate our journey entire...

ACHILLES: And what might that be?

TORTOISE: What if the two of us, this moment, are once again but two characters in some narrative upon a page, strange-looping our way into the mind of yet another reader? What if we are now gone, immortalized in other minds, and are ourselves simply bits and pieces? Could we ever even know it?

ACHILLES: You mean, we have journeyed from (a conversation with Einstein's brain) to (a conversation with the mind that conceived (a conversation with Einstein's brain)) into (a conversation in the mind of someone who strange-looped (a conversation with the mind that conceived (a conversation with Einstein's brain))). A staggering, sobering thought! I suppose, if that were true, we would indeed have no way of knowing.

TORTOISE: Our fate, then, originated in Hofstadter's hands -

ACHILLES: -Lewis Carroll's, I think!

TORTOISE: Stipulated. It started there, passed through Hofstadter, and now resides in the strange loops of the individual who may be reading our words at this very moment. We can only hope we make a favorable impression!

ACHILLES: One hopes indeed!

TORTOISE: And with that, old friend, I suggest we stretch our legs!

(They rise from their café table, enjoy a moment of bright Italian sun, and stroll into the Biblioteca di Brera.)

The Children of Babel:
In the Hall of
All Possible Thoughts

Daniel Dennett is a child of Babel.

And between himself, Hofstadter and Searle, he is probably the one who spends the most time thinking about thinking.

We've already noted his 'intuition pumps' - short, simple thought experiments that stimulate the imagination – and can immediately appreciate Dennett's sophisticated contemplation of what goes on in a conscious mind. An avowed physicalist/cognitivist, he views thought as process, and processes are subject to methodical reduction.

It is with Dennett in mind that we now enter a new quadrant of Babel – a place where we consider a final feature of all possible minds: the thoughts within them, and how they are embodied in Babel.

What does a *thought* look like in Babel?

In approaching that question, we must first settle on a definition of thought. Here are some candidates:

Per Oxford, a thought is "an idea or opinion produced by thinking, or occurring suddenly in the mind," or "the action or process of thinking." Per Meridian-Webster, a thought is "an individual act or product of thinking," or "something (such as an opinion or belief) in the mind." Per dictionary.com, a thought is "the product of mental activity; that which one thinks," or "a consideration or reflection," or "meditation, contemplation, or recollection."

This sampling works well for our purposes, as it demonstrates the considerable range of mental events that qualify as thought. And while there is variation in the definitions, they all are described the same way in

Babel: as lengthy strings encoding the cortical activity in a brain across a certain period of time.

Within this context, we can imagine even more:

- *A developing thought*: a string encoding an idea being pondered from germination to full realization;
- *A strange loop thought*: a string encoding a thought that leads to another, and to another, and to another, and back to the original thought;
- *An irrational thought*: a string encoding a thought that cannot be reconciled with previous thoughts or experience through reason;
- *A continuous thought*: a string encoding a thought that slowly fades into another, and into another, and into another – stream of consciousness.

And there are, of course, many more.

Even with this expanding typology, the Babel strings encoding these thoughts all have some things in common.

1. They are not discrete; they do not begin and end emphatically (every thought leads to another thought);
2. They are constantly evolving, deflected in new directions by interaction with others, environmental input, or physiological changes;
3. As cortical events, they all occur within the perimeter of conscious awareness.

What does this mean in Babel terms? A box set that defines the entire life of a human brain, from birth to death, is in fact one long string, composed of tens of millions of substrings that flow one into another. The transition from one thought to another is necessarily fuzzy; the neural activity being encoded is a continuous stream, persisting even (at much lower levels) during sleep. The brain never stops, and neither does the cortical activity that constitutes thoughts.

All of this is in direct contrast to digital computers, wherein all activity has an explicit start and an explicit termination, and where all activity routinely stops altogether.

HAL 9000 and his sister Sal, both of whom we'd say are conscious despite their digital nature, can serve as our intuition pumps in pondering these dynamics.

These conscious machines are never turned off; they are continuously operational, as human brains are. The key moving part in the plot of *2001:*

A Space Odyssey is Hal's discovery, through his covert surveillance of astronauts Bowman and Poole, that they plan to shut him down. The thought of such an event leads Hal to murder Poole and try to kill Bowman, acts of self-preservation; and when Bowman succeeds in slowly shutting Hal down, the computer pleads for him to stop, openly displaying terror at the prospect of his thoughts terminating.

Almost a decade later, sister Sal is told by Dr. Chandra that he would like to recreate Hal's shutdown for research purposes – he wants to disconnect her higher processing functions, then reconnect them. Both the circumstances and the relationship between human and computer differ in this exchange, and Sal is not terrified. She simply has a question: *Will I dream?*

Of course, Chandra replies. *All intelligent creatures dream. No one knows why.*

We can extract considerable analogia from these scenes. Certainly the prospect of having our brains shut down terrifies us all; brain death is, after all, what death really is. Hal is aware that if his thoughts end, he will end, even if his "body" persists. We all know it's the same for us.

Sal, on the other hand, is approached differently; Chandra carefully (and graciously) explains his intent, doing his utmost to dissuade her from fear at the prospect of disconnection. What he plans to do is likened to sleep, inspiring her question about dreaming.

So it is with us; when we sleep – when we "lose consciousness" - our stream of thoughts continues, our cortical activity persists, but without the guidance of self-awareness and with minimal sensory input. When we are able to remember and examine our dreams, we can trace the stream of consciousness, the one-thought-leads-to-another phenomenon now punctuated with randomness – a faint echo of our waking consciousness.

The point is simple: consciousness is a continuum, not a set of discrete events, and we are at least passively aware of it. We perceive that continuum, to the point of instinctively dreading its interruption.

We can add *continuum* to our list of inherent properties. Consciousness is inherently continuous.

This thought leads to *that* thought, along a narrow path

What follows from this inherent continuity, and how would it be realized in a conscious machine?

First, let's look at how Hofstadter himself defines a "thought" - or rather, how the character Achilles does:

"A thought occurs (in the mind *or* the brain, whichever you prefer – *for now!*) when a series of connected neurons fire in succession – mind you, it may not be a long string of *individual* neurons firing like a chain of dominoes falling down one after another – it may be more like *several* neurons at a time tending to trigger another few, and so forth. More likely than not, some stray neural chains will get started along the side of the mainstream but soon will peter out, as threshold currents are not attained. Thus, one will have, in sum, a broad or narrow squad of firing neurons, transmitting their energy to others in turn, thus forming a dynamic chain that meanders within the brain – its course determined by the various resistances in the axons that are encountered along the way..."[51]

Written 40 years ago, this explanation of thought is both spot-on and a bit naïve (everywhere Achilles says "few" or "several", one should read "thousands" or "tens of thousands"), but it gets us where we're headed next.

The "neural chains" Hofstadter describes – neural networks – are indeed webs of axons connecting neurons by way of synapses, and "thought" is indeed what we call the collective activation of groups of neurons. Elsewhere, Hofstadter terms this activation a "neural flash", "swooping and careening" through the brain - a picturesque and useful descriptor. It evokes an image of the motion of thought, a continuum, and gives us more besides. It articulates that when one thought morphs into another, it is by way of physical connection between neurons.

Put another way, the neural flash can only proceed along neural connections; it cannot impact neurons that aren't in a particular group or chain. Put yet another way, there's a limited range of candidates for *That* thought which can possibly follow *This* thought.

And whatever the current thought in mind might be, it can only have been preceded by thoughts emerging from the activation of neurons connected via axons rearwards in the group or chain.

There are further implications. This architecture of the brain implies that some thoughts are not, and can never be, adjacent; we must proceed from Thought A to Thought B to Thought C in order to get to Thought D, because of the physical connections between them and the potentiation of the relevant synapses.

Except...

Analogia – the creation of analogies – creates shortcuts between thoughts. *This* is like *That* repotentiates synapses, creating frictionless

---

[51] In "A Conversation with Einstein's Brain", *The Mind's I*, p. 436.

pathways between neuron groups. I see a *dog*, I think of a *fire hydrant*. I see a picture of *Nixon*, I think *crook*, and so on. This is truly a defining property of intelligence, and by extension, consciousness: the shortcutting of thoughts in the brain is of tremendous evolutionary advantage, speeding our reaction time, deepening our understanding of our environment (and one another), spreading our knowledge of one domain into others. And the architecture of the brain makes clear that it's not a learned skill; it's built into us.

Some thoughts are not possible in some brains

As an aside, we should make note of the following: the human brain is the most complex object in the known universe – three pounds of protein with literally trillions of moving parts. And here's the wonder of it: my brain has more in common with your brain than any other two objects in the universe have with one another – and yet, *at the same time*, they have more *differences* than any other two objects in the universe!

We are all very, *very* alike. And yet we are all very, *very* different.

We all have around 90 billion neurons, and 4-5 orders of magnitude that number in connections between them. They are *generally* connected the same way, brain region to brain region, but my individual neurons are connected *uniquely*, and any one neuron could have hundreds more or fewer connections than a comparable neuron in your brain in the same brain region.

That means that my "neural flashes" will be different from yours – from almost everyone's. When two people think the same thing at the same time, it owes to identical stimulus in the environment or astronomical chance.

Some analogies don't work for everyone, because they are simply wired differently; some thoughts (or ideas) are beyond some people, because their neuron chains are built differently.

Think about that, the next time you're facing your cranky, opinionated, chronically wrong uncle at the Thanksgiving table.

And there's another factor to consider: when one thought morphs into another, it follows the axon path of optimum potentiation – or, put another way, the path of least resistance. This means we have *very limited control* over our stream of consciousness. We learn to deal with this as children, structuring our play, our tasks, and our interactions in such a way as to keep our "neural flash" more or less on a productive course; but even as mature adults, our brains are constantly seeking out that path of least resistance – and that's the mechanism by which our minds wander.

It need not be so for a conscious AI.

Our own brains follow a common heuristic: neurons fire (pass the "flash" along) when they build up enough energy from the firings of preceding neurons. Every neuron has an energy threshold at which it fires (in neural networks, this threshold is configurable). The path-of-least-resistance rule, then, is the governing principle of the neural activity that generates our thoughts.

In a conscious AI, however, different heuristics could be built in: imagine, for example, a neural network composed of neurons that fired across the path of *most* resistance – making the least-accessible next thought the *most* accessible. Imagine a network that purposefully mixed and matched the two, seeking out the odd thought intentionally, then integrating it cleanly with a root thought.

And it need not be least/most; where the heuristics governing neuron firings are concerned, the sky's the limit; it is already the case that many different methods exist in simulated neural networks, serving all sorts of unusual applications. If we purposefully build heuristics less limited than our own into actual neural networks instantiating machine consciousness, the possibilities are literally beyond imagination.

The frogs and the stream

What does all this look like in the *pi* substrings of Babel?

The box-set string that defines the life of a brain from birth to death, to begin with, isn't all thoughts: it includes the algorithmic messaging from eye and ear to muscle; it includes all the "stray neural chains" Achilles mentioned, which never go anywhere; it includes the noise generated by the senses, which is sometimes relevant but often not; it involves random firings that emerge from internal glitches, triggered by vitamin deficiency or indigestion or fatigue.

We can call the sum of all that content *consciousness*. Within that consciousness, we have *thoughts*, which flow across our self-aware attention; and we sometimes have *expressions of our thoughts*, speaking them or writing them down and submitting them to the self-aware attention of others. *The expression of thought is not the same thing as thought* – different parts of the brain are involved – and this distinction brings us to the finish line in our exploration of the potential for machine consciousness.

Having reached the end, let's strange-loop back to the beginning, and bring in some analogia:

Thoughts are to *consciousness* as *frogs* are to a *stream.*

Thoughts exist, continuously, within the stream of consciousness; they are made of the stuff of consciousness; they grow and morph and interact with other thoughts.

Frogs exist, continuously, within an actual stream; they are made of the stuff of the stream. They begin as one thing and become another, growing, interacting with other creatures – subsuming some, being subsumed by others.

And what about words?

Expressions of thoughts are *frogs jumping out of the stream.*

Expressions of thoughts leave the individual through acts of speech or writing; they continue to exist, apart from their brain of origin; they make their way into other brains, there to live or die or be broken down for their raw materials; sometimes they go nowhere, dying in the air.

Frogs leave the stream and migrate to other streams, or to ponds, etc.; they exist apart from their stream of origin; they interact with the environment beyond the stream, sometimes consuming and sometimes consumed.

Why is he telling us this?

I'm telling you this because one of the key failings of AI research – and the academic enterprise of cognitive science overall – has been an out-of-kilter focus on the frog beyond the stream. Too many contributors to the field have made the *output* of intelligence and consciousness the end-all, be-all, without admitting that only the smallest fraction of our lives as conscious beings involves what we say, what we do, what we write down. Behavior itself is only a product of a fraction of our neural activity; our inner lives are much deeper, much richer than our patterns of movement in the world.

Cognitive science, and AI in particular, have been myopic endeavors, treating communication and behavior as if they *were* consciousness, rather than *products* of consciousness. If we want to produce true machine consciousness, however, we can't simply conjure up frogs – that's what chatbots are. We need a stream, one that includes the building blocks to make tadpoles. That's how we produce real frogs.

But that analogy is silly. Let's try another:

Thoughts are to *consciousness* as a *play* is to a *stage*.

Expressions of thoughts are *the actors on the stage*.

This one might be slightly clunkier than its predecessor, but let's see if it's serviceable. If *thoughts* are the *play*, then all the components of the play – the sets, the lights, the props, the costumes – are connected, each triggering the next in turn, constantly flowing and coalescing – just as thoughts do in consciousness. If the *expressions of thoughts* are *the actors on the stage*, then the performance itself – the actors speaking and interacting – are releasing the story from the stage, presenting it to the audience – just as spoken thoughts make their way into other minds.

Would we say *the actors* are *the play*? Of course not. There is much more to a play than just the actors.

And our final point: a play can be staged in many theaters. It is fallacious to say that only one kind of venue will do; Shakespeare has been performed in traditional theaters, parks, high school auditoriums and church basements. To be Shakespeare, there must of course be many ingredients – sets and props and costumes and, certainly, actors and words – but any venue that can accommodate the ingredients will do.

Even a Chinese Room.

Defining Requirements of Consciousness

Analogical Thought

Strange Loops

Meaning

Experience

Intentionality

Distributed Processing

Parallel Processing

Mobility

Community

Self-Awareness

World Modeling

Theory of Mind

Intuition

Continuum

Entropy, Interrupted:
Consciousness and Energy

The American physicist Jeremy England has posited a new theory which proposes that the conflict between entropy (the idea that the universe tears down order to create chaos) and life (the continuous build-up of evermore-complicated organic molecules and increasingly complex living organisms) is solved by a simple principle: it is the nature of matter to *spread energy*.[52]

At the most extreme macro level, this makes perfect sense and is observably happening: since the Big Bang, energy has been pouring out into the universe like spilled water on a kitchen floor, finding every possible path to every possible destination. Like kitchen furniture, the specks of matter in the universe - planets, comets, asteroids, interstellar dust - block and deflect that energy. And, as kitchen furniture gets wet, matter in the path of energy soaks it up.

Per Einstein, we know that matter itself (including us!) is simply condensed energy. That condensation is nothing more than an incidental consequence of certain thermodynamic conditions in certain places in space. Our planet, its atmosphere, its water, all the life on it - including us! - are an incidental mass of condensed energy, no different than all the other chunks of not-quite-energy dancing around stars and wandering through cold interstellar space.

Except...

Here on this particular planet, things are a little bit different.

---

[52] The first part of this essay originally appeared in the collection *I Think I'm Right in Saying That?* (2019), by the author.

Spreading Energy

First, a bit more of England: his team observes in their published papers that the *Spread Energy* imperative, elegant and simple, is observable everywhere in Nature. So ubiquitous is it, when sought out, that it is clearly the rule, not the exception: the properties of matter, both living and non-, all give service to this imperative. The *Spread Energy* imperative appears to be the very nature of Nature (though the theory is not yet proven).

England appears as a character in Dan Brown's novel *Origin*, which provides examples of natural organization that promotes entropy: simple objects, like snowflakes - frozen water - which spontaneously form complex shapes to most efficiently distribute light and heat. They cannot *not* do so. Quartz displays similar properties – elegant organization of non-living matter, optimized to radiate energy efficiently.

Origin's examples include weather, which can be characterized as complex systems that optimize to dissipate energy in the atmosphere, from the pressure-relieving vortex of a tornado to the electrical discharge of a lightning bolt, dispelling the structured ganging of charged particles in a thundercloud.

Even the simple mechanisms of life demonstrate England's principle – and life turns out to be matter's best innovation yet, when it comes to the spreading of energy. Photosynthesis, *Origin* points out, is a marvelous example: a tree absorbs a steady stream of sunlight, absconding with some of it to extend and replicate itself, while dissipating the remainder as infrared radiation: increased entropy.

And as for DNA, the engine of organic replication – it, too, exists in the service of entropy: a forest, for instance, can dissipate far more energy than a single tree.

If all of this is how Nature really works - and it appears increasingly likely - then England's Spread Energy principle explains how organic molecules came to be, and why increasing complexity is their rule: what we call life is in fact just another expression of matter's inherent imperative to get out of energy's way, and to exploit energy in that endeavor.

(This new feature of the universe, by the way, closes the final gap that God has been filling since Darwin. With the tendency of organic molecules to form under certain thermodynamic conditions, and for those molecules to form increasingly complex structures, the origin of life is now explained in full - the need for a 'Creator' has finally expired. The question, "How do living systems arise from non-living matter?" has finally been answered, if England is correct. This is a major theme of *Origin*.)

Physicists have already skipped ahead to the end of the universe's book: in the last pages of the final chapter, it will experience 'heat death' - the final cessation of all energy exchange, as there will eventually be no remaining thermodynamic fuel for entropic processes. England's model not only supports this long-accepted conclusion - it helps explain it.

Between now and the death of the universe, per England, the energy that it contains will continue to spread relentlessly - and all the specks of condensation, all the matter, will inadvertently interrupt that spread, catching and trapping energy in the process. And that captured energy will cause the substance of that matter, the cold particles that cause it to be, to rally for its release, scrambling to combine with other particles in whatever way will promote the energy.

There is no other outcome. All of physics, all of reality, all the laws of Nature yet discovered, bow in service to this unceasing agenda.

...including, again, us.[53]

The Last Resort

It is matter's job to get out of energy's way, however it can, as fast as it can. It is energy's job, when it attempts to spread and finds itself thwarted by intervening matter, to infiltrate that matter and do what it must to push through; and, if possible, cause that matter to return to its own natural state – to *become* energy, when conditions are favorable.

If England's inspiration causes us to reconsider the origins and purpose of life, how does that inform our opinions about whether there's any more of it out there in the universe?

If anything, the England Imperative - *Spread Energy!* - makes the universe an easier place to understand. We are already very clear on the laws of thermodynamics, and so the imperative isn't exactly a sharp turn. But it casts a new light on our assumption that since we live in a universe where the elements of life are abundant, given uncounted billions of opportunities to occur, life will certainly make many appearances.

---

[53] Credit where it's due: the idea of dissipative adaptation, proposed and promoted by England, owes much to the work of Nobel laureate Ilya Prigogine, a Russian physical chemist, who proposed Classical Irreversible Thermodynamics, a mathematical formulation that attributed the self-organization of matter as chemical potential. His work, in turn, was based on that of American physical chemist Lars Onsager, who likewise won the Nobel Prize in Chemistry.

The England Imperative strongly suggests that life is only going to arise in places where matter can't easily spread energy in a less complicated way. Energy, interrupted by matter, will push through the path of least resistance; matter, organizing to optimize energy's quest, will only become as complex as it needs to be to accomplish that mission. Put another way, life – as the most complex of matter's mechanisms for dissipating energy – is a last resort.

A wickedly simple expression of the England Imperative might be an asteroid orbiting a star - composed of pure iron, let's say, and spinning slowly. The area of the asteroid facing the star will absorb energy from the star, then shed it into cold space as it rotates. It is, at most, a very temporary interruption of energy's journey to where. Nothing as intricate as life is necessary for the asteroid to perform its essential energy hand-off.

Even if the asteroid contained the elements of life, it wouldn't matter: because the thermodynamic system of the asteroid is so simple and performs so efficiently, nothing else is needed - and the additional conditions we know to be necessary for life are absent, in any case.

But an amazing thing happens when we begin to add those conditions: we begin to see direct parallels between the conditions required for life and the complexity of the energy capture of the place where it can take root. Put another way, a celestial body of great thermodynamic complexity and a planet that has conditions friendly to the emergence of life are one and the same.

The more thermodynamically complex the planet, the greater its friendliness to the emergence of life.

Using Earth as our only real baseline, we can observe that even with its staggeringly complex thermodynamic systems, life took its sweet time developing here. Once it did, it began doing its job - *spreading energy!* - and it is no great undertaking to chart the changes in the thermodynamics of the Earth in parallel with the increasing complexity of its emerging biosphere. What had to happen for such a system to emerge was for the planet's thermodynamic complexity to occur in the first place.

Let's examine that complexity briefly.

Complexities

The most famous of the characteristics of a life-bearing planet is that it be a Goldilocks planet. This means the planet exists in its parent star's *circumstellar habitable zone* (CHZ) - not too close, not too far away - so that its surface can support liquid water. The not-too-close/not-too-far, then, translates to not-too-hot/not-too-cold, in terms of surface temperature.

To have liquid water on its surface, the planet must have atmospheric pressure to hold it in place - and to retain an atmosphere heavy enough for that kind of pressure, it must be impervious to solar winds, which would ordinarily strip away the atmosphere of a planet so close to its star. Earth passes this test (while, for instance, Mars does not), possessing a rotating liquid nickel-iron core (a natural dynamo) that gives the planet a solar radiation-repelling magnetic field. [Venus possesses only a sparse magnetic field, yet holds an atmosphere far denser than Earth's; even so, the consequences of this are life-prohibitive - the solar winds are indeed eroding its upper atmosphere, and that activity has been depleting the planet of low-mass hydrogen and oxygen ions for billions of years, ridding it of all the water it may have originally possessed.]

A planet must also rotate at a certain pace - fast enough to pass heat in the course of rotation such that the surface water doesn't ever get hot enough to boil outright, but slowly enough that some of it evaporates, in order to carry water to land by way of clouds. And its mass must be such that it creates a gravitational field wherein life, once it forms, is secure without being crushed.

These characteristics, in combination, put a planet in an entirely different thermodynamic class than a simple iron asteroid. As the planet rotates, it will catch and release heat, as the asteroid does - but with liquid water on its surface and a gentle thermal cycle, vast amounts of energy will nonetheless be trapped to endlessly cycle between land, sea, and air.

The endless cycling of energy creates for Earth an England problem: the matter contained within this huge thermal vista must further organize and complexify in the service of spreading it. Snowflakes happen; tornadoes spring up.

But even that isn't enough, thermodynamically, to create a need for an energy-spreading mechanism as complex as life.

The Earth traps energy above and beyond storing it in air and water and rocks: it shakes and stirs it like a bartender.

Earth spins on an axis, as all planets do. But the spin of the Earth literally has a twist: the planet's axis is tilted, rather than perpendicular to the solar plane. This is, of course, the basis of the seasons. And the seasons are, almost by definition, the gradual shifting of energy from one planetary region to another, in yet another endless thermal cycle, as it revolves around the sun.

And the shaking and stirring goes further still: even more thermodynamically important than the Earth's axial tilt, perhaps, is the gravitational effects of its oversized moon. At one-quarter the size of the Earth itself, the moon is by far the largest satellite of a planet known to exist, ratio-wise.

The consequence, of course, is the intense tidal forces it brings: the moon churns the planet's oceans, generating still more energy to be trapped in liquid matter. The oceans are literally vast reservoirs of energy.

Now the complexity of the Earth has it working overtime to spread energy, because there are so many constantly-interacting systems trapping and re-trapping energy within - and new mechanisms for spreading it must rally to the task. Elements must combine, molecules become more intricate. New systems must interrupt the existing ones, to push against the bulwark of this tense equilibrium.

Now Life emerges. Self-replicating energy couriers of microscopic size, dedicated to scooping up energy from the environment and radiating it out, with far more intensity than a snowflake and far more intricacy than a tornado. A whole new chapter begins, as life interrupts not only the equilibrium of Earth's dancing thermodynamic systems, but their chemical composition as well. Carbon dioxide, methane, and other heat-trapping molecules of the sort that are currently broiling Venus become essential players in this disruption, paradoxically leveraged to solve the problem they themselves create. If Earth was a complicated story before, now it's a James Joyce novel.

And the supreme complexity in the system – life! – serves to introduce new complications that perpetuate the precarious balance between thermodynamic systems, even as they succeed in spreading energy to high heaven. We have named those complications *flora* and *fauna*.

In its simplest forms, life is a marvelous dissipative adaptation; even in an incarnation as simple as algae, it absorbs and redistributes energy, using some of it to replicate itself and thereby increase its utility, thus contributing to entropy. And as we've noted above, more sophisticated forms - land-based plant life being a great example - are even more impactful, turning an empty expanse of heat-absorbing dirt into a sprawling, infrared-radiant entropy engine that will spread and spread.

Algae and plants - all forms of life, really – are reorganizations of matter in the service of energy redistribution. But unlike snowflakes and tornadoes and other transitory, one-off mechanisms, living organisms themselves take on the role of energy trap: they collect more energy than their raw materials would, if dissociated, and apply that energy to the purposes of entropy.

And when we get to animal life, the England Imperative gets more creative still: animals don't just absorb solar energy, exploit it, then redistribute it; animals can capture and trap the energy *of other living things*, other energy traps. They extract the energy of other plants and animals. In this process, entropy is served on yet another level: the animal consumption of other forms of life, in releasing that life's energy, breaks it down - *order into chaos* - entropy, once again.

This stupendous innovation is remarkable enough in itself. But it gives rise to one of the most unusual features of the Earth's biosphere: the thermodynamically precarious swap of gases between plant life and animal life. Respiration is experienced by plants and animals alike, with plants emitting oxygen and animals emitting carbon dioxide - a complementary accommodation that benefits both, keeping both kingdoms of life in equilibrium. If either kingdom were to vanish from the Earth, the other would be hard-pressed to survive.

And it grows more complicated still: as animal life has taken the lead as both the premiere redistributor of energy on the planet and its most explicit and dedicated hoarder, it simultaneously (and unwittingly) contributes to the planet's entropy interruption: carbon dioxide and methane, the two gases animals emit, are themselves energy traps. They capture and retain heat; when expelled into the atmosphere, they cause the atmosphere itself to hoard more energy than before.

Repurposed

In hindsight, the emergence of life here on Earth seems inevitable; but the take-home point is that the Earth is as thermodynamically different even from planets of comparable size as an internal combustion engine is from a wind-up watch. We are children of the most unlikely of planets. To say that Earth is one in a billion would be almost certainly understate. And, as a side note, the implications of the England Imperative make the ubiquity of complex life in the universe a *very* distant possibility. It is highly improbable that the universe is teeming with life – which makes the life here on Earth all the more precious.

We can, if England's theory proves true, employ it as a predictive tool as we venture out into the universe: encountering new worlds, we would estimate the likelihood of finding life or something like it based on the world's observable energy traps, its thermodynamic complexity. The more straightforward its energy transfer, the less need for an energy dissipation system as intricate as life.

Here on Earth itself, however, it would have been impossible, given the planet's staggering energy burden, for life *not* to have emerged: our planet is a clogged energy sink, capturing the radiation of the sun and hoarding it shamelessly, passing it from system to system, releasing to the night only what it absolutely can't permanently ensnare during the day.

The conclusion is simple and profound and deeply disturbing: we are products of entropy interruption. And our purpose, as far as the universe is concerned, is the spreading of energy. We are here only because in this universe, elements combine in the way that most efficiently pushes energy along, and our particular molecules happen to be situated in the mother of all energy traps.

Any purpose beyond that transcends that of the universe - and is entirely ours to define.

Stars are the forges of energy, emitting it endlessly in an effort to warm the void; planets are interrupters of that entropy; and we are, ironically, planetary disrupters.

We are both agents of and disruptors of entropy. We, the highest form of life, having emerged as reality's champions of energy dissipation, are uniquely positioned to seize the reins of energy distribution – and have already begun doing so, in the order we have created for ourselves out of the raw materials of the Earth. As self-directed entropy engines, we can use our power to increase the interruption of entropy in the service of order, redirecting the dissipation of energy in a manner that complements our industry as the gases of plants and animals complement one another. We can refine our powers of disruption, redirect the role of matter, and leverage entropy's chaos as a means, not an end. We can, put simply, repurpose the universe.

Origin, indeed...

Ruminations

Dissipative adaptation - 'spreading energy' - may, in the long run, turn out to be an insight into the nature of matter on a par with quantum mechanics. If true, it closes once and for all the gaps in our understanding of our origins, and gives us a completely new perspective on how reality is constructed.

But what does it have to do with consciousness?

If organic matter was a dissipative adaptation that led to reproductive flora and fauna, as England's work suggests, then life imbued with consciousness may be the next adaptation beyond that. That is, if we grant England's argument that matter self-organized as reproducing life is more efficient in energy distribution than inorganic matter, then we can ask if conscious organisms represent an improvement on non-conscious organisms, when it comes to the spreading of energy.

Intuitively, it's obvious that this is in fact the case: examples of human beings harnessing and distributing energy are all around us, and have been throughout the history of civilization, from the construction of buildings to the cultivation of plants; from the systematic herding and consumption of animals to the use of fire to manipulate metal; from the industrial burning of fossil fuels to the harnessing of solar and nuclear power. We are *masters* of efficient energy distribution.

On the other hand, we are also a species that is, through its collective, conscious activity, constructing energy traps – and one in particular, the modification of the planet's atmosphere. Its radical imbalancing, a consequence of our industrial activity, is trapping more and more energy, increasing risk not just to ourselves but to the planet's ecosystem as a whole. We, as a conscious, intelligent species, make a greater gross energy distribution contribution than any other – but as we work *against* the spreading of energy by trapping it, our net contribution is falling. We can then wonder if another adaptation must now present itself, to re-optimize the weakened equilibrium.

Brown doesn't say it this way, but his follow-up to the presentation of England's thesis in *Origin* may answer that question. We've already mentioned it in a previous chapter: *Technium*.

The hybridization of humankind and the technology we've produced into a new species may – almost certainly will – occur, and when it does, it will clearly play a role in whatever happens next with our tendency, as energy agents, to work against ourselves. There is no clear solution to the vast energy trap we've created, no clear path to reducing the threat to ourselves we've produced that can achieve the necessary scale. On the other hand, Augmented Intelligence will necessarily open doors we haven't even yet imagined, let alone attempted to pass through; we can anticipate that our successor species will not only come up with solutions we don't yet envision, but produce the new technologies they require.

Moreover, our hybrid descendants will have tools we didn't have when we started creating this mess: systems that can watch over the natural world, learn more about it than we ever could, applying that knowledge to intervene and fine-tune it passively, and not only restore but improve upon the equilibria we've so ignorantly disrupted. The more conscious neohominins of kingdom *Technium* will easily surpass us as the most efficient energy distributors that have ever existed.

We shall see.

The Requirements and Properties of Consciousness

Analogical Thought

Strange Loops

Meaning

Experience

Intentionality

Mobility

Community

Self-Awareness

World Modeling

Theory of Mind

Intuition

Spreading Energy

Conclusions

I spent my graduate school years peeking into the brains of newborn infants and young children. We were using a gentle EEG technique to measure the responses of developing brains to certain sounds (most often human vocalizations) and occasionally the presentation of stimulating images. We were building on a method of detecting incipient learning deficits, the idea being that early detection would lead to early intervention.

This experience, combined with the literature I had to absorb at the time, gave me a deep appreciation of the connection between the physical brain and mental states. I had minored in philosophy as an undergrad, and had come away from those heady seminars with a smorgasbord of impressions about mental states; and the data we scooped out of those babies and gradeschoolers made very clear that most of those impressions were far afield of reality. Descartes was flying in heavy clouds.

The preceding essays may seem, to some degree, frivolous, and I openly concede that they are not as well-integrated as their arrangement would imply; but my intent is not to persuade, or even made distinct claims – it has been an exercise in strange-looping, introducing bits of my own consciousness to the reader for consideration, à la Rod Serling. I have found the works of Hofstadter, Dennett, and Searle (and many others in the bibliography to follow) fascinating, stimulating, and certainly consciousness-expanding, and it is my hope that exposure to these ideas might trigger similar reactions.

Even so, there are claims to be made, culled not from any one source above, but from the confluence of them all:

The hardware matters.

In addition to the neuroscience and philosophy education mentioned above, I've spent my adult life immersed in computer science, more than 20 years of it as a technologist. I've written thousands of programs, hundreds of applications, designed dozens of architectures, even designed control systems for custom robotic hardware for the Department of Defense. And I can state unequivocally that the brain-computer analogy is nonsense.

Your mind is the software, your brain is the hardware, that's how its advocates often express it. The implication is that our physical brains and our thoughts, our consciousness, are distinct and separable – that minds are something apart from brains. And this leads to all sorts of corollaries, such as the notion that our minds can be uploaded from our brains into computers.

Anyone who believes this doesn't really understand either brains or computers.

Computers (which we created not to replicate our brains but to compensate for their deficiencies) have given us some misguided notions, primarily the restoration of Descartes' error – the idea that the mind and brain are separable. But the more deeply we investigate the brain, and the more broadly we understand how computers work, the weaker this misguided notion becomes: human consciousness emerges, not from the combination of hardware and software, but from wetware – hardware that is its own software. Refuting the analogy with itself, we can say that in the brain, the "programming" is inseparable from the "processor".

Psychologist Stephen Kosslyn, former Dean of Social Science at Harvard University:

"Mental capacities such as memory, perception, mental imagery, language, and thought all have proven to have complex underlying structures. Cognitive neuroscientists improve our understanding of them by delineating component processes and specifying the way they work together.

"Researchers in cognitive psychology and some parts of artificial intelligence share this aim, but they do not consider the brain. Their central metaphor is the computer. Just as information processing operations in a computer can be analyzed without regard for the physical machine itself, mental events can be examined without regard for the brain. This approach is like understanding the properties and uses of a building independently of the materials used to construct it; the shapes and functions of rooms, windows, arches, and so forth can be discussed without reference to whether the building is made of wood, brick, or stone. We call this approach *Dry Mind*.

"In contrast, we call the approach of cognitive neuroscience *Wet Mind*. This approach capitalizes on the idea that *the mind is what the brain does*: a description of mental events is a description of brain function, and facts about the brain are needed to characterize these events.

"The aim is not to replace a description of mental events by a description of brain activity. This would be like replacing a description of architecture with a description of building materials. Although the nature of the materials restricts the kinds of buildings that can be built, it does not characterize their function or design. Nevertheless, the kinds of designs that are feasible depend on the nature of the materials. Skyscrapers cannot be built with only boards and nails, and minds do not arise from just any material substrate."

Details matter, Dennett said in one of his Chinese Room screeds[54]. Yes indeed, they surely do! But if they matter in the Chinese Room, how much more so in the brain itself? It is disingenuous to apply such rigor to a thought experiment, then dismiss it entirely in the real world; the hardware "doesn't matter?" It surely does, and we now live in a time when any competent application developer or systems architect will tell us so: the idea that "software" can run on "any hardware" is naïve, and betrays a lack of knowledge of how complex information systems actually work.

Once again, let's agree that just because minds and brains aren't "software" and "hardware" *does not mean* that we can't create artificial minds! We simply need to consider different approaches, and "wetware" gives us a big leap forward.

We underscore the earlier distinction between a simulated neural network and an actual one: they produce the same output, but they aren't the same thing at all. If we want to truly define consciousness in such a way that we can create it artificially, we need to look beneath the output and understand the processes.

Simulation is just that.

Stepping away from the idea that the output of consciousness is all there is to consciousness, we likewise step away from the idea that the

---

[54] In *Intuition Pumps*, p. 325.

simulation of a thing is the thing itself. It's impossible to make that mistake in John Searle's example, the one pointing out that a perfect simulation of a rainstorm by a computer won't leave anybody wet; it's easy to do when the output looks and sounds and feels like us, and our Theory of Mind switch is being hammered.

We have Alan Turing to blame for this one, I think; the idea that disembodied human conversation might cause that "community" component of consciousness to come alive isn't factored into his Test. It wasn't meant to explicate consciousness, but to support an abstract claim connecting universal information processing paradigms to the idea of non-biological intelligence. It has misled us terribly, but the fact that the Turing Test (which was developed decades before *any* of the discussions above were even possible) addresses only one of many components of consciousness shouldn't distract us from the bigger suggestion that artificial consciousness is possible; we simply have to be more inclusive about what that artificial consciousness would contain.

If I create a robotic simulation of myself that can perfectly reproduce my expressions of thought – and even my behaviors, the way I walk, my facial expressions, etc. - I have created something that even my closest friends and family might mistake for me. That's an impressive accomplishment; but it's not the same as reproducing my own consciousness. The latter contains myriad internal processes and experiences and sensations that never make it into my words or actions – and they are not only a part of my consciousness, they are a greater proportion of it than the words and action are.

And this leads us to our next claim...

Consciousness is far more than what we say and do.

Once we get past the outsized focus on a system's output, accepting that what lies beneath the surface is integral to consciousness, we need to figure out what it is beneath the surface that matters.

Does this mean that an artificial consciousness must reproduce not only human output, but human sub-surface processes?

That's an important question. Because these things *aren't* observable via verbal expression or behavior, but only through intuition and self-report, their recreation is a far greater challenge than producing systems that pass the Turing Test.

But there are a few pieces we do have, and they're mentioned above. Strange loops, for one; continuous streams of thought, for another. These, we know are there; I've seen them myself, observing the brainwaves of children. On and on they go, and when the child produces an explicit response to an explicit stimulus, suddenly the brainwaves reorganize, very briefly – and then they return to their sub-surface cascade, energetic but inscrutable.

And we can safely claim that for an artificial system to be conscious, it must be self-aware; it must experience things. That requires some active mechanism, working in parallel with the stream of thought, observing its own operations and associating those observations with past events and perceptions and sensations. Must that, too, mimic the human version?

There is no reason to think so. Those components are necessary for the artificial consciousness to satisfy our definitions, but there is no reason in principle that they cannot be achieved in non-biological media. Put another way, an artificial consciousness must be more than its output, it must have an "inner life" to qualify as conscious; but that inner life can be utterly alien, from a human point of view, and still be consciousness. It is anthropological arrogance to suppose otherwise.

Consciousness is a group activity.

It can't be overstated that strange loops are central to consciousness; it may be that they, more than any other component we've considered, *define* consciousness.

There is irony in the fact that Hofstadter eschews the idea of consciousness, calling it an illusion, citing strange loops as evidence. That's a matter of perspective, but it doesn't detract from the fact that the *I* that defines me as a conscious being is a product of my endless interaction with other *I*s. We get our *I* from the group.

So, necessarily, must any artificial consciousness. A self-awareness mechanism is essential, but alone is insufficient; to build a *self* requires raw materials – experience – and that experience must include interaction with other consciousnesses. Hal must have Frank and Dave, or he cannot form a distinct identity.

We waited more than 50 years for the advent of artificial intelligence, and now it's upon us. After decades of pontifications and assumptions and suppositions and disturbing summer blockbusters, all of which served to misinform and distract us and warp our understanding, AI has quietly slipped into our lives, already so pervasive and entrenched that we'll never be without it. It is simultaneously invasive and unobtrusive, popping up in unwanted Internet ads while weaving itself into our healthcare and financial systems, unobserved.

The AI that has stolen into our institutions and our offices and our homes and our pockets is nothing like what we imagined. It has marched past the sentries that have stood watch since the Sixties – in academia, in industry, even in pop culture – and made itself completely at home, without their permission. It has nested itself in technology along the path of least resistance (profitability), effortlessly toppling the couch-cushion forts of modern philosophy along the way. Most of what the old-guard punditry has said or claimed about it over the intervening decades has turned out to be wrong or irrelevant.

Now the moment we've anticipated for more than two generations is upon us. Intelligence machines are a daily reality, and they are not only improving everything they touch, they are beginning to improve *us* – augmenting our insights, our performance, our predictions in ways we would never have anticipated, even a decade ago. They will continue to do so, in ways we've yet to imagine.

But even this isn't the pinnacle; *machine consciousness* still beckons. Some think it's only a couple of decades away, others think centuries. Some say never.

One thing is certain: when it arrives, it will be as its precursor has been – a thief in the night, slipping past our pontifications and assumptions and suppositions and defying all our expectations. We need to get ready for it, to adjust our thinking, start paying attention, and seeking out a new and broader perspective. AI is already changing how we think, how we do things.

Per Douglas, Daniel, and John, machine consciousness will change who we are – and who we can become.

Bibliography/ Recommended Reading

2001: A Space Odyssey, Arthur C. Clarke. New American Library, 1968.

2010: Odyssey Two, Arthur C. Clarke. Rosetta Books, 2012.

The Astonishing Hypothesis: The Scientific Search for the Soul, Francis Crick. Touchstone, 1995.

The Children of Babel: Essays on the Inherent Nature of Artificial Intelligence and Consciousness, Scott Robinson. Paleos Media, 2020.

A Companion to Cognitive Science (Blackwell Companions to Philosophy), ed. William Bechtel and George Graham. Wiley-Blackwell, 1998.

A Companion to the Philosophy of Mind (Blackwell Companions to Philosophy), ed. Samuel Guttenplan. John Wiley & Sons, 1996.

The Conscious Mind: In Search of a Fundamental Theory, David Chalmers. Oxford University Press, 1997).

Consciousness and Language, John Searle. Cambridge University Press, 2002.

Consciousness Explained, Daniel Dennett. Back Bay Books, 1992.

Contact, Carl Sagan. Simon & Schuster, 1985.

Darwin's Dangerous Idea, Daniel Dennett. Simon & Schuster, 1996.

The Engine of Reason, The Seat of the Soul: A Philosophical Journey into the Brain, Paul Churchland. MIT Press, 1995.

Fantasia Mathematica, Clifton Fadiman, ed. Copernicus, 1997.

Fluid Concepts and Creative Analogies: Computer Models of the Fundamental Mechanisms of Thought, Douglas Hofstadter. Basic Books, 1996.

Gödel, Escher, Bach: An Eternal Golden Braid, Douglas Hofstadter. Basic Books (anniversary edition), 1999.

HAL 9000: An Unauthorized Biography, Scott Robinson, Paleos Media, 2020.

HAL's Legacy: 2001's Computer as Dream and Reality, ed. David G. Stork. MIT Press, 1997.

I Am a Strange Loop, Douglas Hofstadter. Basic Books, 2007.

I Think I'm Right in Saying That?, Scott Robinson. Paleos Media, 2019.

Infinity and the Mind: The Science and Philosophy of the Infinite, Rudy Rucker. Birkhauser, 1982.

The Intentional Stance, Daniel Dennett. Bradford Books, 1989.

Intentionality, John Searle. Cambridge University Press, 1983.

Intuition Pumps and Other Tools for Thinking, Daniel Dennett. W.W. Norton & Company, 2014.

The Joy of π, David Blatner. Walker Books, 1997.

Labyrinths, Jorge Luis Borges. New Directions, 2007.

Le Ton Beau de Marot: In Praise of the Music of Language, Douglas Hofstadter. Basic Books, 1997

Macmillan Dictionary of Psychology, Stuart Sutherland. Palgrave MacMillan, 1991.

Mappings in Thought and Language, Gilles Fauconnier. Cambridge University Press, 1997.

The Master and His Emissary, Iain McGilchrist. Yale University Press, 2019.

Mind-Body Problems: Science, Subjectivity & Who We Really Are, John Horgan. Knot Press, 2019.

The Mind's I: Fantasies and Reflections on Self and Soul, Douglas Hofstadter, Daniel Dennett. Basic Books, 1981.

Minds, Brains and Science, John Searle. Harvard University Press, 1984.

The Mystery of Consciousness, John Searle. The New York Review of Books, 1997.

Neuroscience and Philosophy: Brain, Mind, and Language, Daniel Robinson et al. Columbia University Press, 2009.

On Intelligence: How a New Understanding of the Brain Will Lead to the Creation of Truly Intelligent Machines, Jeff Hawkins and Sandra Blakeslee. St. Martin's Griffin, 2005.

Order Out of Chaos: Man's New Dialogue with Nature, Ilya Prigogine and Isabelle Stengers. William Heinemann Ltd., 1984.

Organization of Behavior: A Neuropsychological Theory, Donald Hebb. Wiley, 1949.

Origin, Dan Brown. Transworld Publishers, 2018.

The Origin of Consciousness in the Breakdown of the Bicameral Mind, Julian Jaynes. Mariner Books, 2000.

Quiddities: An Intermittently Philosophical Dictionary, W.V. Quine. Belknap Press, 1989.

Routledge Encyclopedia of Philosophy, ed. Edward Craig. Routledge, 1998.

The Society of Mind, Marvin Minsky. Simon & Schuster, 1988.

Stairway to the Mind, Alwyn Scott. Copernicus, 1999.

Surfaces and Essences: Analogy as the Fuel and Fire of Thinking, Douglas Hofstadter and Emmanuel Sander.

The Stuff of Thought: Language as a Window into Human Nature, Steven Pinker. Penguin Books, 2008.

Time Enough for Love: The Lives of Lazarus Long, Robert Heinlein. Ace, 1998.

Views of the Chinese Room: New Essays on Searle and Artificial Intelligence, ed. John Preston and Mark Bishop. Clarendon Press, 2002.

Westworld and Philosophy: If You Go Looking for the Truth, Get the Whole Thing (The Blackwell Philosophy and Pop Culture Series), ed. James B. South and Kimberly S. Engels. Wiley, 2018.

Wet Mind, Stephen M. Kosslyn and Oliver Koenig. Simon & Schuster, 1995.

About the author

Scott Robinson is a social scientist and technologist, specializing in AI and its positive social and operational applications in the enterprise. His background includes both academic research in social psychology and technology development, and he is currently engaged in research and development of Augmented Intelligence. He can be reached at scottrobinsonwriter@gmail.com.

www.ingramcontent.com/pod-product-compliance
Lightning Source LLC
Chambersburg PA
CBHW070634220526
45466CB00001B/174